MANGO: Preventive Practices *and* Curative Measures

MANGO
Preventive Practices and Curative Measures

H.S. SINGH, VISHAL NATH
ABHA SINGH & S. MANDAL

2008

SSPH

SATISH SERIAL PUBLISHING HOUSE
403, Express Tower, Commercial Complex, Azadpur, Delhi-110033 (India)
Phone : 011-27672469, 27672852, Fax : 91-11-27672046
E-mail :info@satishserial.com, hkjain1975@yahoo.com
Website : www.satishserial.com

Published by :

SATISH SERIAL PUBLISHING HOUSE
403, Express Tower, Commercial Complex, Azadpur, Delhi - 110033 (INDIA)
Phone : 011-27672852 Fax : 91-11-27672046
E-mail : info@satishserial.com, hkjain1975@yahoo.com

ISBN 81-89304-55-0

Printed at
SALASAR IMAGING SYSTEMS, DELHI-35

ABOUT THE AUTHOR

Dr. Hari Shankar Singh obtained Doctorate degree in Agricultural Entomology from N.D. University of Agriculture and Technology,Faizabad. He served in Central Silk Board in Jammu and Kashmir as Senior Research Associate, Indira Gandhi Krishi Viswa Vidyalaya, Raipur, Chattisgarh as scientist, Vivekanand Parvatiya Krishi Anusandhan Sansthan –ICAR, Almora-Uttaranchal and Horticulture and Agro-forestry Programme, Ranchi-Jharkhand as Sr. Scientist. He has published nearly 40 research papers, 35 articles, many bulletins, abstracts and pamphlets. He has handled and worked in the projects funded by NHB, APCESS, DFID, NATP and DBT in the field of insect pest management in fruits and vegetable crops. At present, he is serving as Senior Scientist at Central Horticultural Experiment Station, Bhubaneswar- Orissa, under Indian Institute of Horticultural Research-ICAR.

Prof. Vishal Nath, has completed his Master's degree from Narendra Dev University of Agricultural and Technology, Faizabad and Ph.D. from Kanpur. He worked as scientist (Horticulture) at Central Institute for Sub-tropical Horticulture, Lucknow and NRC for Arid Horticulture, Bikaner. He served as Senior Scientist at Central Horticultural Experiment Station (IIHR), Ranchi and joined as Professor (Horticulture) at Birsa Agricultural University, Ranchi Presently, he is working as Principal Scientist & Head, Central Horticultural Experiment Station (IIHR), Bhubaneswar. Apart from his contribution to National Agricultural Research System (NARS), he has wider exposure to Hi-tech Horticulture being practiced in developed countries like Israel. Prof. Vishal Nath has exclusively worked on fruits of major and minor importance being grown in arid and semi-arid regions and under waste land conditions. He has developed many technologies and varieties of different fruit crops suitable for resource poor areas. He has written several technical bulletins, extension folders and information brochures on these aspects. Prof. Vishal Nath has authored very important books on aird horticulture, Litchi and Fruit Production Technology for Plateau regions. Prof. Vishal Nath has more than 200 publications on different fruits which includes research papers, popular article, technical bulletins, research manuals and information bulletins in his credit.

Smt. Abha Singh obtained her master degree from Banaras Hindu University and joined ICAR as scientist. Basically she is involved in research and training in the field of food and nutrition with special reference to fruit and vegetable preservation. She has developed low cost vitamin A rich sweet potato based weaning mix for children. She has published 10 research papers, 18 articles, in addition to abstracts and popular articles. Presently she is working as scientist in National Research Centre for Woman In Agriculture, Bhubaneswar –Orissa.

Shri Sudhamoy Mandal graduated from Bidhan Chandra Krishi Viswavidyalaya (BCKV), Mohanpur and obtained his Masters degree in Plant Pathology from Indian Agricultural Research Institute (IARI), New Delhi. He was awarded Junior Research Fellowship of Indian Council of Agricultural Research (ICAR), New Delhi during M. Sc. study. He joined Agricultural Research Service (ARS) as a Scientist (Plant Pathology) at Central Horticultural Experiment Station, Bhubaneswar. His research interests include biological control of plant diseases, management of fruit and vegetable diseases using agro-residues and systemic acquired resistance (SAR). He is on the verge of completion of his Ph. D. from Indian Institute of Technology Kharagpur (IIT-Kgp).

PREFACE

India is the largest producer of mango in the world. Andhra Pradesh, Uttar Pradesh, Bihar, Maharastra, Karnataka, Kerala and Tamil Nadu are the major mango producing states. Area under mango is further expected to increase in many other states after implementation of National Horticulture Mission. Dashehari, Langra, Banganpalli, Alphonso, Kesar, Totapuri, Himsagar and some hybrids are considered among the most superior types of mangoes in India. Commercial mango farming is done by large farmers; however, the scenario is changing gradually due to participation of small and marginal farmers. The productivity of mango is low due to poor initial planning, inferior planting material, poor orchard management practices, unscientific post harvest practices and natural calamities. Vegetative dominance over reproductive phase, irregular flowering, erratic fruit set and its growth, fruit drop, post harvest losses, pests, diseases, disorders and adverse climatic conditions affect the economic cultivation of mango. In order to avoid loss and increasing profitability, it is necessary to take steps from initial stage so that vigour and growth of the orchard is maintained by performing proper operations right from planning of orchard to sale of the produce.

There are two types of losses in mango cultivation. First arising due to manmade mistakes and carelessness such as poor planning, poor orchard management and post harvest handling and second, arising due to other factors such as pests, diseases, disorders and adverse climatic conditions. Manmade mistakes may occur in setting up of an orchard, selecting varieties, planting, fertilizing, drainage, irrigation systems, shaping, weed control, mulching, pruning and canopy management, use of growth regulators, harvesting, ripening, packing, marketing and quality management. Pests, diseases and fruit disorders may occur by their own or may be aggravated further due to negligence of farmers in orchard planning and management. In mango cultivation, today major issues are nutrition, irrigation, pests, diseases and disorders which can be rated as the critical issues for attention to improve its productivity and quality. Improvement in production of planting material is an area that needs action to stop or slow down the re-cycling of disease and pest problems in old and new orchards.

Educating growers, contractors and extension officers on these aspects to broaden their knowledge base is the key issue of the day. The present book has been prepared to address the field related problems of mango growers in action mode with a view that scientific planning, good horticultural practices and proper pest knowledge save a farmer from problems and losses in future. The book is presented in simple language without the load of conflicting views so that readers can understand and take proper steps from very beginning to avoid problems in future. Alltogether, 21 chapters and an appendix of the book cover various aspects of mango cultivation and problems, based on updated information vis-à-vis the research and field experiences of the authors'.

We are thankful to all the technology sources and persons, whose findings and views have been incorporated in this book. We extend our gratitude to our colleagues, Dr.G. Naik, Dr. D. Kumar and Dr. V. Pandey and other staff members of Central Horticultural Experiment Station for their co operation in preparation of this manuscript.We are specially thankful to Sri.K.G. S. Rao,computer technician,CHES for his valuable help in the preparation of this manuscript.

We hope that beginners as well as large number of traditional mango farmers will be immensely benefited from this publication.

25.2.2008 **H. S. SINGH**

CONTENT

Preface v

1. **Introduction** **1-7**

2. **Understand The Growth and Fruiting Behaviour of Mango** **9-16**

    Growth Stages of Mango, 9

    Nursery stage, 9

    Establishment and non-bearing stage, 9

    Junior adult and adult bearing stage, 10

    Senior adult or full bearing stage, 10

    Ageing stage, 11

    Year Round Growth and Fruiting Pattern, 11

    Flushing or emergence of new shoots, 11

    Dormancy or rest period, 11

    Flowering and Fruiting, 12

    Flowering, 12

    Fruit set, 13

    Fruit drop, 14

    Fruit growth and development, 14

    Fruiting season and yield, 15

3. **Start with Scientific Orchard Planning to Avoid Future Problems** **17-35**

    Suitable Site and Soil, 17

    Select suitable soil type, 18

    Soil inherent nutritional status, 19

    Abnormal soil pH, 20

    Water Table, 20

Suitable Climate, 21
Altitude, 21
Temperature regime, 21
Rain fall pattern and humidity, 22
Frost and hail storm, 23
High wind and cyclone, 23
Selection of Varieties, 25
Varieties of different regions, 27
Maturity Period, 29
Select a Few Among Many Varieties, 29
Hybrids, 33

4. Learn to Produce and Use the Quality Planting Material 37-46

Nursery Act, 38
Selection of Site, 38
Selection of Varieties, 39
Establishment of Nursery Structure/Materials, 39
Development of Mango Scion Block, 40
Preparation of Rootstock, 40
Planting in Polyethylene Pots or Planting a Secondary Nursery, 41
Selection of Scion, 42
Propagation Techniques, 42
Mannuring, Watering, Drainage and Plant Protection, 46

5. Adopt Standard Orchard Establishment Practices 47-62

Land Preparation and Appropriate Layout, 47
Standard Size Pit and its Filling, 49
Select Healthy and Genuine Saplings, 50
Buyers guide for mango saplings, 51
Hardening the Sapling Before Planting, 53
Ensure Minimum Root Disturbance of Grafts During Transport and Planting, 53
Scientific Planting Technique, 54
Appropriate time, 54
Proper distance, 55

Right method, 56
Irrigation, 56
Apply Mulch and Provide Shade, 57
Make Provision to Drain Excess Water, 57
Plant Live Fence and Shelter Belt as Wind Break, 57
Importance of Live Fencing and Wind Break, 58
Disadvantages of Live Fences and Windbreak, 59
How does a windbreak work? 60
Live fence posts, 61
Live barriers or hedges, 61

6. Use *in-situ* Orchard Establishment Technique in Hilly, Sloppy & Adverse Land conditions 63-69

Land Development, 63
Orchard Layout, 64
Pit Digging and Filling, 64
In situ Planting of Rootstock, 64
In situ Grafting, 65
Care and Management of Plants, 66
Canopy Management, 66
Utilization of the Interspaces for Cultivation, 67
Saving the Young Plant from Enemies, Pests and Fire, 67
Soil Erosion and Water Management on Sloping Orchards, 67
Hillside Ditch, 68
Vegetative Barriers and Conservation Cover, 68
Mulching and Residue Management, 68
Irrigation Practices, 68

7. Maximize Profit Through High Density Planting and Orchard Floor Management 71-82

High Density Planting, 71
Reasons for HDP as Against Normal Planting, 72
Major components of HDP, 73
Planting density, 73
Planting system, 73
Pruning system, 74
Automatic systems, 75
Nutrition and irrigation, 75

Control of Pests and Diseases, 76
Fruit Yield, 76
Economic impact, 76
Orchard floor management, 76
Consideration for Intercropping and Cover Cropping in Mango, 77
Method of growing, 81
Pest management, 81
Irrigation and nutrition, 81

8. Take Special Care and Manage the Developing Plants 83-88

Do not Negelect Plant Basins, 83
Remove Wild Sprouts and Flower Buds, 84
Protect Plants from Cold and Hot Air, 84
Proper Water Management Practices, 85
Balanced Fertilizers, 86
Weed Management, 86
Mulch in Plant Basins, 87
Initial Framing, 87
Inter Crop and Cover Crop, 88
Bio-Fencing and Wind Break, 88
Protection from Insect Pests and Fire, 88

9. Ensure that Plant is Nutritionally Balanced 89-97

Some Points to Consider in Fertilizer Management in Mango, 89
Role of Major Nutrients in Mango, 90
Nutrition Management, 90
Nursery stage, 91
Transplanting stage, 91
Initial bearing and bearing stages, 92
Method of fertilizer application, 93
Foliar feeding, 93
Micronutrients Deficiency and Correction, 94
Recommended Concentrations of Micro Nutrients for Foliar Sprays, 97

10. Avoid Moisture Stress and Weed Excess at Crtitical Stages 99-110

Factors to be considered for optimum irrigation in mango, 100

Soil and climatic factors, 100

Root depth and its expansion, 100

Stage of Plant Growth Decides Irrigation, 101

Precautions/Considerations in Mango Irrigation, 101

Irrigation in bearing orchard, 102

Method of irrigation, 103

Weed Management Practices, 106

Methods of weed management, 106

Mannual, mechnical and chemical practices, 106

Intercropping/cover Cropping for Weed Management, 107

Mulching for weed management and moisture conservation, 107

Different types of mulches, 109

Organic mulches, 109

Inorganic mulches, 110

Method of mulch application, 110

11. Manage the Canopy and Rejuvenate the Old Trees 111-124

Importance of Ideal mango Canopy, 111

Proper Planning for Canopy Management, 113

Methods of Framing of Newly Planted Orchard, 115

Maximizing Production in Bearing Trees Through Pruning, 117

Restructuring Canopy in Big Trees, 119

Top Working for Rejuvenation, 120

Economic Impact, 123

Basic Pruning Tools, 123

Other Operations, 123

12. Realise Full Potential of Bearing Orchard 125-132

Nutrient Management Practice, 126

Fertilizer Placement, 129

Lime Application, 129

Irrigate the Plant as Per Recommendation, 129

Increasing Pollination, 131
Pruning After Harvest, 131
Thinning out Excess Fruits, 132
Management of Tree Basin, 132
Pest and Disease Management, 132

13. Make Mango Orchard to Bear Every Year 133-147

Factors of Irregular Bearing, 134
- Irregular bearing variety, 134
- Growth pattern of mango, 135
- Unfavorable climate, 135
- Nitrogen and carbohydrate reserves, 136
- Irregular bearing may be due to problems in planting site, 137
- Hormonal imbalance and physiology of shoots, 137
- Poor tree management, 138
- Additional nutrient may not help, 138
- Excessive crop load, 139

Management of Irregular Bearing, 140
- Cultural management, 140
  - Deblossoming and thinning of fruits, 140
  - Pruning, 141
  - Controlled irrigation and manuring, 142
  - Smudging, 142
  - Others, 143
- Use of Chemicals, 143
  - Plant growth regulators, 143
  - Paclobutrazol (PBZ), 143
- Use of potassium nitrate, 146

14. Understand the Role of Pest and Pesticide Management in Mango Health 149-158

Key Aspects and Concept of IPM, 149
Methods of IPM, 151
Cultural Practices, 151
Biological Factors, 153
Chemical Factors, 153
Proper Care and Handling of Pesticides, 154
At the Time of Purchase, 154

Storage and Disposal, 156
Routes of Entry of Pesticide, 156
Personal Protective Equipment (PPE), 157

15. Adopt Preventive Practices and Curative Options for Insect Pests Management 159-192

Pests of Mango Fruit and Their Management, 160
Fruit Flies (*Bactrocera species*), 160
Management, 161
Bait application technique (BAT), 161
Spot Application of BAT, 162
Precaution, 162
Male Annihilation Technique (MAT), 162
Method, 163
Chemical Control, 165
Cultural, 165
Other Methods, 166
Integrated Control, 166
Post-Harvest Disinfestations, 167
Fruit Borer (*Deonalis albizonalis*), 168
Management, 168
Mango Stone Weevil *(Sternochetus mangiferae*), 169
Fruit Sucking Moths (*Othries* sp., *Achaea* sp.), 171
Pests of Mango Flower, 173
Mango Hoppers *(Idioscopus sp. and Amritodus atkinsoni), 173*
Mango Gall Midges, 175
Pests of Mango Buds and Leaves, 177
Leaf Cutting Weevil (*Deporaus marginatus*), 177
Mango Mealybug, 177
Leaf Eating Caterpillars, 179
Shoot Borer (*Chlumetia transversa*), 180
Leaf Miner *(Acrocercops syngramm), 181*
Aphids, 182
Flea Beetle (*Rhynchaenus mangiferae*), 182
Grey Weevil (*Myllocerus* species), 183
Mango Shoot Gall Psylla *(Apsylla cistellata), 183*
Thrips, 184

Leaf Gall Midge (*Procontarinia matteriana*), 184
Red Tree Ant (*Oecophylla smaragdina*), 185
Mango Scale Insects / Coccids, 185
Black Fly (*Aleurocanthus woglumi), 187*
Tea Mosquito Bug (*Helopeltis antonii), 187*
Mango Leaf Webbers (Orthaga species), 188
Mites (*Paratetranychus mangiferus, P.yothersi.*
and *Eriophyes* sp.), 189
Pests of Mango Trunks, Twigs and Roots, 190
Bark Eating Caterpillars *(Inderbela species), 190*
Stem Borer *(Batocera rufomaculata)*, 191

16. Adopt Preventive Practices and Curative Options for Disease Management 193-207
Fungal Diseases, 194
Bacterial Diseases, 204
Algal Diseases, 205
Phanerogamic Parasite, 206

17. Adopt Preventive Practices and Curative Options for the Management of Disorders 209-222
Malformation, 209
Black tip, 212
Soft nose, 213
Internal necrosis, 213
Clustering of Fruits, 214
Spongy tissue and leaf scorching, 215
Fruit drop, 216
Stem End cavity, 217
Sun scald (Heat Injury), 217
Chilling Injury, 218
Bark cracking, 218
Woody gall, sap burn and scarified fruits, 219
Skin abrasion and lenticels spots, 220
Little leaf and copper deficiency symptoms, 220
Frost infury, jelly seed, 221
Black pimples and fruit tumors, 222

18. Adopt Monthly Work Schedule to Keep Mango Orchard Healthy and Productive 223-230

19. Reduce Harvest, Handling and Marketing Losses 231-250

Ensuring Right Stage of Harvesting, 232
Maturity Indices, 232
Dipping Test, 233
Planning for Harvesting, 235
Trainging to Workers, 236
Picking Time, Techniques and Materials, 237
Harvesting Season of Mango in the Country, 240
Desapping, 240
Handling from Field to Pack House, 241
Final Grading and Ripening Aids, 242
Packing and Transport, 246
Storage and Marketing, 248
Price structure, 250

20. Extend Self life, Reduce Post Harvest Losses and Convert Into Value Added Products 251-267

Post harvest life of mangoes, 251
Methods to increase the post harvest life, 252
Pre-harvest tratment, 252
Post-harvest reatment, 253
Converting into value added products, 254
Product and market potential, 255
Raw mango produts, 256
Ripe mango products, 261
Ripe mango Powder, 263
Mango Jam, 263
Mango Slices, 265
Mango Juice, 265
Mango Leather and bar, 266
Mango Cereal Flakes, 266
Mango Squash, 267

21. Identify the Problems and Select Appropriate Remedy as and When Needed 269-276

22. Appendix 277-285

1. *Trichoderma enrichment of FYM, 277*
2. *Methods of leaf sampling for nutrient analysis, 277*
3. *Method of soil sampling for nutritional analysis, 279*
4. *Preparation and use of Bordeaux mixture, 280*
5. *Bordeaux Paste, 282*
6. *Bordeaux Paint, 282*
7. *Important Mango Hybrids and their Parentage, 283*
8. *Technology Sources, 284*
9. *Cost of Mango Cultivation, 285*

23. Bibliography 287-292

24. Colour Plates 293-301

1

Introduction

Many farmers in our couuntry, now -a -days, wish to invest in mango as this crop has ready market and high yield and demand in local and export market thereby, good profit. Traditionally, it is a favourite fruit crop of India and can be easily grown in suitable climate. It contributes 40 per cent to the total fruit production in the country and occupies 42 per cent of the total area under fruit crops and about 41 per cent of world mango production (Table-1). Uttar Pradesh and Andhra Pradesh have the highest acreage under mango followed by Bihar, Maharastra, Karnataka, Kerala and Tamil Nadu. Indian mangoes are known for their taste and flavour and rule the domestic market as king of fruits during the season. They have great potential for export also and have major demand in Middle East countries.A limited quantity is also exported to European market; however, its export share is limited to 15 per cent, that too of varieties like Alphonso, Dashehari and Kesar etc. Today, the mango is cultivated in almost all the states of India and some growing pockets in different state have been marked and declared as mango belt (Table-2) for commercial production. In these areas, regulatory norms have been fixed to provide safe environment and logistic support system is being developed to have better production and marketing as special zone.

General mango cultivation as well as trade oriented mango production is preference oriented, for a specific type and flavour. By and large there are mainly two types of mangoes i.e. monoembryonic or varieties of subtropical origin and polyembryonic or varieties of

Table 1. Top ten mango producing countries

Country	Mango production (tonnes)	Per cent
India	10,800,000	41.00
China	3,582,000	13.00
Thailand	1,700,000	6.00
Mexico	1,503,010	6.00
Pakistan	1,089,000	4.00
Philippines	967,535	4.00
Brazil	850,000	3.00
Nigeria	730,000	3.00
Egypt	327,000	1.00
Australia	60,000	0.24

Source: FAO 2004 and Mango Industry Strategic Plan 04, Australlia

Table 2. Mango growing pockets of country

State	Growing Belts
Andhra Pradesh	Krishna, East and West Godavari,Vishakhapatnam, Srikakulam, Chittoor, Adilabad, Khamman, Vijaynagaram
Chhattisgarh	Raipur, Bastar, Durg, Bilaspur, Bastar, Rajnandgaon, Rajgiri
Gujarat	Bhavnagar, Surat, Valsad, Junagarh, Mehsana, Khera
Haryana	Karnal, Kurushetra
Jammu & Kashmir	Jammu, Kathuwa, Udhampur
Jharkhand	Ranchi, Simdega, Gumla, Hazaribagh, Dumka, Sahibganj, Godda
Karnataka	Kolar, Bangalore, Tumkur, Kadagu
Kerala	Kannur, Palakkad, Trissur, Malappuram
Madhya Pradesh	Rewa, Satna, Jabalpur, Katni, Balaghat
Maharashtra	Ratnagiri, Sindhudurg, Raigarh

Contd...

Table 2. Contd...

Orissa	Sonepur, Bolangir, Gajapati, Koraput, Rayagada, Gunpur, Malkangiri, Dhenkanal, Ganjam, Puri
Punjab	Gurdaspur, Hoshiarpur, Ropar
Tamil Nadu	Dharmapuri, Vellore, Tiruvallur, Theni, Madurai
Uttaranchal	Almora, Nainital, Dehradun, Bageshwar, Udhamsingh Nagar, Haridwar
Uttar Pradesh	Saharanpur, Bulandshahar, Lucknow, Faizabad Barabanki, Varanasi
West Bengal	Malda, Murshidabad, Nadia

tropical origin. Mango is a cross pollinated crop and new varieties have been evolved as a result of natural and man made crosses within and between above said two groups. Over the years, these varieties have moved from place to place around the world and acclimatized for cultivation. Thus, a range of widely differing varieties has appeared over the last few decades. Out of these varying populations, superior genotypes producing high yields and quality fruits have been adopted as cultivars (cultivated varieties) in different mango growing regions. Most of these have been mass propagated through vegetative means to maintain their genetic purity. These cultivars over the years have become an important component of the liking of the common man and have established as commercial cultivars. These cultivars are location specific and have distinct behaviour in relation to bearing habits, fruit size, color, flavor, aroma and yields. On the other hand, they are very sensitive to the growing environment. Therefore, any change in environmental factors, severely affects the productivity and fruit quality of these cultivars. This is the reason why a cultivar has its domain at one place not at the other.

Mango fruit, unripe / ripe is consumed as fresh and in various value added farms. Pickles and chutneys are made from unripe mango. Different types of drinks, jams, jellies are prepared using ripe fruit. Likewise, dried and dehydrated mango pieces and mango powder are also seen in the market. Both, unripe and ripe mangoes have medicinal properties. A drink made out of unripe mango is used as a remedy to prevent various body ailments caused by a raise in ambient air temperature. Unripe mangoes are also used in treating stomach problems and to stimulate bile formation and in treatment of blood related diseases. Consumption of ripe mango is useful to overcome night blindness and to protect health of skin. There is a common belief that consumption of ripe mango with cow's milk helps in weight gain.

The level of various nutrients in mango may vary depending on the cultivar, ripeness of the fruit and area of cultivation. The normal nutritional value of the mango has been given in table -3.

Table 3. Nutritional composition of mango*

S.No.	Matter/nutrient	Nutrient level /100g of mango flesh
1	Water	81.0 g
2	Energy	74 k cal
3	Protein	0.6 g
4	Lipids	0.4 g
5	Carbohydrates	16.9 g
6	Calcium	14 mg
7	Phosphorus	16 mg
8	Iron	1.3 mg
9	Carotene	2743 micro g
10	Thiamin	80 micro g
11	Riboflavin	90 micro g
12	Niacin	0.9 mg
13	Vitamin C	16 mg

*National Institute of Nutrition

Mango seeds, leaves and bark are used in the treatment of diarrhea and disorders in reproductive system of women. Dried leaves are used for mulching tree basins or as fuel. The seed kernel is generally eaten during famine period with or without processing. In areas of hilly Orissa; preparations from the kernel powder are consumed even if there is no shortage of food. Leaves have allelopathic effect to control weeds and green leaves are sometimes fed to cattle and are used in various medicines. Mango plant, leaves and wood have varied uses in religious, spiritual and domestic purposes. The tree itself is a useful windbreak. These characteristics features make mango tree as an outstanding multipurpose tree.

Commercial mango plantation creates a source of job opportunity for the people associated with growing and managing orchards, harvesting and post harvest handling, packing, transportation, export and value addition. Commercial mango production has attained the status of an industry with forward and backward linkages.

A profitable mango industry depends on the production capacity of orchards. Lack of proper initial planning and poor orchard management practices lead to various kinds of production related problems in the established mango orchard. Trees must be in position to produce sufficient quantity of fruits every year for market balance. Therefore, it is important that the trees are managed with scientific tree management system. Besides, the crop growing conditions, the cultural practices and the cultivar determine the quality of mango at a particular location. Mineral deficiencies, improper water and nutrient management, poor pest management and cultural neglect eventually affect the yield and quality. The following factors may cause future problem and loss (Fig. 1) to any farmer having a mango plantation, or planning to venture into mango industry.

- Poor understanding of mango growth pattern and behavior leading to unscientific plant, pest and produce management.
- Mango cultivation under unsuitable geographical location and unsuitable cultivars under poor growing conditions
- Poor planning, such as unsuitable spacing (high density / regular), faulty layout, poor pit preparation, pit mixture and pit filling, improper production or procurement of planting material,odd time planting, post planting neglect, faulty tree frame and fruiting arms etc.
- No adoption of proper orchard management techniques like tree basin and orchard floor management, pruning, weeding, methods for ensuring regular bearing orchard neglect causing ill tree health, undesirable water and nutrient management methods and schedule.
- Poor skill of pest and disease identification and their management.
- Poor knowledge about harvesting time and method, production volume and prices at farm gate, wholesale and retail markets. Unscientific post harvest handlings at field level and at packing house.
- Poor efforts to increase shelf life of fruits and preserved products.

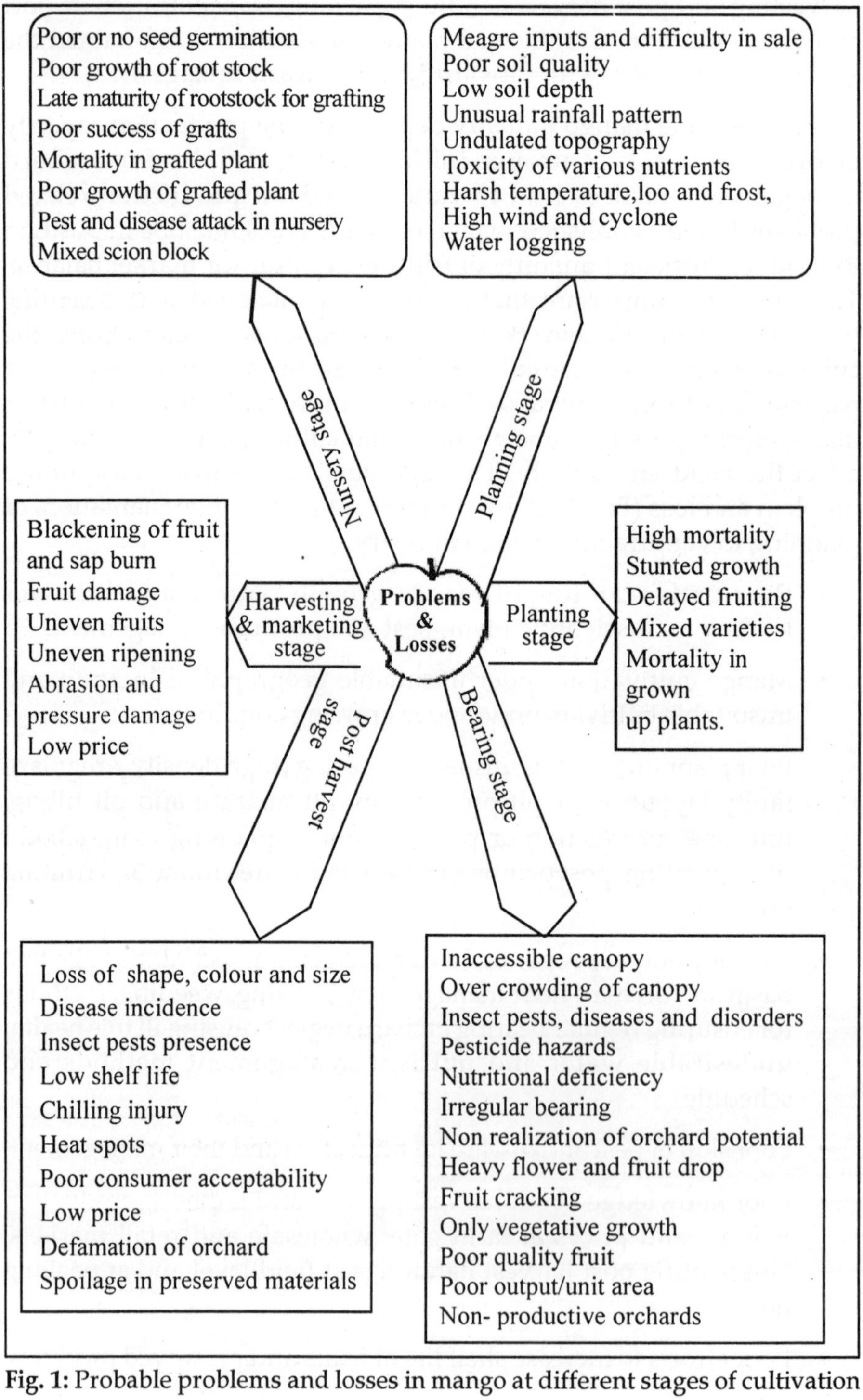

Fig. 1: Probable problems and losses in mango at different stages of cultivation

- Poor knowledge about methods to withstand the post-harvest abuses of the marketing and distribution process as markets can accept 'only limited quantities'.
- Unawareness of pest clinic information services, directory information services, policy information services and bazaar information services.

An effort has been made to develop this book perceiving the "good horticultural practices as tools for prevention of problems and losses in later stage"in mango production. Attempts have also been made to enlight the readers regarding preventive practices that can be adopted to avoid the losses and curative options that can mitigate the problems.The information furnished in the manuscript is based on the working experience,survey of available literature and applied aspects of core research.

□□□

2

Understand the Growth and Fruiting Behaviour of Mango

Before under taking commercial plantation of mango, one should understand that how mango plant grows over the years and yields fruit because this basic information makes strong base for initial planning and applied knowledge for orchard management. The orchard management practices differ with the age and stage of mango plant. The practices for juvenile plant (non bearing young plant) would be different from the senile (bearing) plant. Broadly, based on progressive age, the mango plant has 6 distinct phases of growth and development. Based upon these stages, its behaviour may vary and so the management practices.

Growth Stages of Mango

Nursery stage

The very first stage of mango is known as nursery stage between 0-1 years. During this stage the plants are maintained by expert nursery operators.

Establisment and non-bearing stage

The second stage lies between 1-3 years. It is the young non-bearing stage in which the plant tries to attain the desired frame. During this stage, plant has accelerated growth to develop scaffolds and main limbs and produces vegetative flushes 2-3 times/year. Due to senility (tendency to bear früit) in the scion (grafted shoot), some plants produce panicles which should not be allowed. In this phase, the orchardist should start efforts to develop uniform frame of the tree. At the terminal end of this stage, some panicles can be allowed to bear fruits but the number should be regulated depending upon the health of the tree. When plant starts bearing, generally two flushes take place, one after fruit harvest and the other during spring. Irrigation and fertilizer requirement of this stage vary from other stages.

Junior adult bearing stage

The third stage of mango plant between 3-5 years of age is known as junior adult bearing stage. This is a delicate stage where a major transformation from vegetative to reproductive growth takes place. During this stage the plant keeps enlarging its canopy and also produces fruits. At this stage, it is essential to maintain a balance between vegetative growth and fruiting since the young plants tends to have more vegetative growth. Some times,the fruit production is suppressed or reduced due to excessive vegetative flushing particularly in spring season during this stage.

Adult bearing stage

Between 6-10 years of age, mango plant enters into a phase where vegetative growth and fruiting remain increasing and that so is percieved during years to come. Balance nutrition of the trees helps in getting consistent yield increase during this period. During this stage, plant produces normally two flushes; 1st during July-August and second during Feb-March. If pruning operation is done immediately after harvest, the 2nd flush comes in form of panicle and the plant produces regular fruits under proper management practices. In some cases, alternate bearing is also seen in later part of this stage.

Senior adult or full bearing stage

The stage of mango tree between 11 to 40 years of age is known as full bearing stage. It has been seen that the growth pattern in mango becomes rhythmic after 20 years. The main features of this stage are well distributed tertiary branches and fruiting terminals which bear the fruits and vegetative flush. At this stage, mango plant shows the tendency of biennial bearing because the plant is exhausted in one crop of 'on year' and takes rest or otherwise the reserve food do not support the fruiting in the next year. To have regular yields, the vegetative flushing pattern, its time and health should be monitored and attempts should be made to check excessive vegetative growth through chemical and cultural manipulations.

Ageing stage

This stage comes 40 years onward symptoms of ageing are evident. The plant becomes old with big canopy and low fruiting. Such tree need special management plan.

Year Round Growth and Fruiting Pattern

Flushing or emergence of new shoots

Flushing refers to the emergence of new shoots on the terminals of old shoots. Major flush takes place just after harvesting of fruits or a little later. The flushing enlarges tree size and produces leaves for utilizing sun light for carbohydrates synthesis which supports fruit development in the following season. Though mango needs only one full flush per year for best yield, usually it has a tendency to produce more flushes than actually required. This is very common when the water and nutrients are not the limiting factors and environmental temperatures are higher than normal for the particular variety. Recurring vegetative flushing modifies the physiology of plant and reduces the yield. Therefore, it is essential to control the growth of shoot flushes to avoid impact on the tree productivity.

Dormancy or rest period

Generally, the mango trees undergo a period of rest or dormancy.It is a period without any externally observable growth activity in mature shoots for about 3 - 5 months depending on the variety. This is a pre condition required for the successful flowering. This dormancy is naturally induced by low temperatures in subtropical environments and soil moisture deficits under tropical conditions or may be by both ways under some situations. Low nitrogen supply also promotes this rest in mature shoots. Though, the term dormancy or rest is used to explain this stage of growth, plants are not physiologically in a dormant stage. During this period plants photosynthesize very actively and produce carbohydrates for storage which is used for further physiological control of flower bud differentiation and flowering. High temperature, high soil moisture and high nitrogen levels in the soil or in plant tissues might break this rest and force the plant to flush again and again. Induced flushing affects the physiology of trees by changing the food reserve base and leads to a further vegetative growth. Plants under such conditions produce poor crops during the ensuing fruiting season. This generally happens due to planting of a cultivar which is not suitable for a particular climate.When a variety, with a tendency to have rest under low temperature climates is planted under high temperature environments, it does not undergo rest and continues to produce vegetative flushes which hampers its performance in the changed environment.

Flowering and Fruiting

Flowering

After the rest period, the mango plant enters into the floriferous phase.During rest period, the level of carbohydrate reserves in plant increases as compared to nitrogen level. Thus, C:N (carbon and nitrogen) ratio of the trees increases.When all the conditions are accomplished and environment is appropriate, the mature shoots become physiologically ready for flowering.

Flowering in mango depends upon bud differentiation in a particular shoot, which generally takes place between September to December, depending upon the local climatic conditions to a larger extent and cultivar to a certain extent. The mango shoot requires certain period for physiological maturity to get differentiated from vegetative primordia to reproductive primordia. It takes about a fortnight for the tiny bud to develop and open into a flower. During the change in primordia, the concentration of Gibberellic Acid (GA_3) either suppressed or the concentration of auxin surpasses the GA_3 concentration. As a result the substance 'florigin' is synthesized in the shoot which induces flower bud differentiation (FBD). Therefore, anti Gibberellin substances (retardants) have been found effective in inducing flowering in some cultivars. The cultivars which produce two flushes due to their genetic make up (cv. Baramasi,) or environmental factors (south Indian conditions) do not require prolonged period for physiological maturity of shoots and can flower twice. Earlier, under north Indian conditions, 5-9 months shoot maturity was required for FBD in traditional cultivars but now cv. Amrapali can bear flowers on 4-5 month old healthy shoots.

The whole process of flowering takes place from December (Andhra Pradesh) to March (Bihar). Flowering may some times start as early as November in East and the West Coast of India. In northern India, mango flowers from February to March with period of full bloom during March.

Under milder climatic conditions of the southern and western India, mango may start flowering from December itself, whereas under extreme climatic conditions of the north, the flowering period is during February- March. Therefore, bloom period in south India is earlier than in the north.

The panicle of the mango varies in length from a few centimeters to 60 cm.Generally, the mango panicle bears two types of flowers i.e. male or staminate flowers and bisexual or hermaphrodite flowers. Complete flowers or hermaphrodite flowers have both male and female organs i.e. anthers and ovary (perfect flowers), where as male flowers have only anthers, producing pollen. The fruit set takes place in only hermaphrodite flowers. There appears to be a close association between high temperature and increased percentage of perfect flowers, and low temperature and decreased percentage of perfect flowers. The percentage of perfect flowers in some of the south Indian varieties has been observed to be much less under north Indian conditions. This has been attributed to the lower maximum and minimum temperatures obtained during the period of panicle development in north compared with those recieved in south India.

The male flower appears first in the panicle and drop off before the opening of hermaphrodite flowers. This situation compels mango for cross pollination. Self unfruitfulness (inability of pollen to fertilize the hermaphrodite flowers of same panicle) and self sterility (inability of pollen of same hermaphrodite flowers to fertilize its own ovary) have also been found in some mango cultivars. As a process, once the hermaphrodite flowers open and the stigma become receptive, pollination and fertilization take place. Insects and house flies are the main pollinating agents. Mango panicles generally produce flowers to the tune of 1,000 to 6,000 depending upon the variety. Out of which, only a few are hermaphrodite and get fertilized ,therefore,only a few fruits or even in some varieties one fruit per panicle starts development whereas in some varieties several fruits per panicle commence further development.

Fruit Set

The initial fruit set depends much on the number of hermaphrodite flowers in a panicle. The percentage of hermaphrodite flowers in a panicle depends upon the early or late emergence of the panicle and the variety. Panicles in the inner portion of the tree bear 1.5-2 times more perfect flowers than those located on the periphery. The number of hermaphrodite flowers is the least in the upper part of the panicle but the percentage of set is the highest. After fertilization, the ovary enlarges and small green fruit sets. The whole process from opening of hermaphrodite flowers and visible fruit set takes 5-7 days. Thus, within 20-25 days, more than 90% fruit set takes place. At this point of time, abrasion in climatic conditions, imposition of stresses and spray of harmful chemicals hamper the process of pollination, fertilization and finally the fruit set.

Soon after fertilization, the plant becomes active in translocating carbohydrate and other food reserves from the storage tissues (source) such as roots and trunks to the developing fruits (sink) for their growth. Fruit set and ultimate retention per panicle are much higher in the medium and late emerged panicles than in the early ones. Fruit set is a varietal character, depending upon several factors such as time of flowering, sex ratio, efficient cross-pollination and intensity of flower drop. Varieties differ from one to other in these respects which leads to varying fruit set in different varieties.

Fruit Drop

There is a heavy drop of hermaphrodite flowers and young fruits, amounting to 99 per cent or more. In the commercial varieties only 13 to 28 per cent of the bisexual flowers have been found to set fruits, out of which only 0.1 to 0.25 per cent reach maturity which indicates the nature of heavy fruit drop in mango. In general, in mango 0.1 per cent or less hermaphrodite flowers develop fruits to maturity. Fruit drop is governed physiologically by the reserve food base in the tree and also by climatic conditions prevailing during this time.

Fruit Growth and Development

The crop load which a tree can bear is primarily determined by the tree reserve level. Excess fruit drop may happen at various stages of fruit development. Even if trees have sufficient reserves, lack of potassium or soil moisture deficit during fruit development phase or pest and disease attacks on flowers and developing fruits may act as limiting factors that control the quantum of the final crop. Therefore, at this stage trees have to be managed very well providing all necessary inputs correctly.

Growth and development in fruit starts after fruit set in sigmoidal pattern.There is a constant increase in skin, pulp and seed of fruits which become stationary at the time of stone hardening and start of qualitative development.In North India, development of fruit starts in the last week of March and is completed by the end of second week of June. The rate of increase in length, breadth and thickness is maximum in April, followed by May, March and June respectively. However, the maximum increase in weight and volume of fruits is recorded in May, followed by April and June, respectively. This indicates that the mango fruit growth and development is sigmoidal in which April month is crucial for fruit size and May for fruit weight. As an orchardist, one should have best care for mango orchards during these months.

Many characters of the pericarp such as cell size, laticiferous canals, intercellular spaces, etc. in different tissues of the fruit contribute towards the increase in length, breadth, thickness and volume of the fruit. Increase in weight is associated with the accumulation of starch grains in the cells. Development of the seed is similar to that of the fruit and thus, the peak growth period of fruit is directly associated with the peak growth period of seed. In mango, the first rapid fruit growth phase is directly governed by the activity of auxin and gibberellins in the seed whereas the second period of rapid development of fruit is due to rapid development of seed and decrease in the inhibitor content of the pericarp.

Finally, in the last stage of the growth and development of fruit, the exocarp region develops into a leathery protective skin (peal), the mesocarp into a fleshy and pulpy region (pulp) and the endocarp into a hard, stony and non-edible region (stone). The growth in the exocarp is due to cell division, increase in cell size and increase in size and number of laticiferous canal.

Fruiting Seasons and Yield

At a particular place mango harvest season limited to 45-60 days only.Based on soil, variety and climatic zones mango fruits mature from February-March (Southern most states) to August (North Indian states). Thus, mango fruits available for about 6 months of the year in various parts. Attempts being made at many centres to extend harvest season.

The yield of mango varies greatly depends upon the variety and agro-climatic conditions prevailing in a region. Grafted mango trees start bearing from the third to fifth year onward. However, seedling trees may take 8-10 years. At the start of bearing, the yield may be as low as 10-15 fruits (2-3 kg) per tree, rising to 50-75 fruits (10-15 kg) in the subsequent years, and to about 500 fruits (100 kg) in its tenth year. In the age group of 20- 40 years, a tree bears 1,000-3,000 fruits (200-600 kg) in an 'on' year. The productive age of a grafted mango tree is usually 40-50 years, after which the yield declines.

❑❑❑

3

Start with Scientific Orchard Planning to Avoid Future Problems

Establishment of mango orchard is a long-term investment as trees continue to produce fruit for longer period than many other fruit crops. Thus, its planting requires proper planning which includes careful selection of site, provision of gentle slope to facilitate proper irrigation and prompt drainage to avoid harmful effect of water logging during rainy season. Therefore, in order to avoid any future problem to appear, special attention is needed in selection of site, field preparation, actual layout of the orchard, proper spacing, digging of pits, selection of planting material and planting, handling of young plants and intercultural operations. Basically, mango is a crop of warmer region, however, it can be grown in cold areas up to 1000m above the mean sea level, but the ideal is 600 msl.At a particular place, the available natural resource based planning should be done to cultivate mango. The mango tree management is specific to variety and plant growth stage but location specific environmental condition do have impact on crop performance. Therefore, the following care should be taken while selecting the area for mango orchard establishment.

Suitable Site and Soil

Mango plantation can only be made commercially viable if it is planted at suitable site. In selecting the site for mango, pre-requisite factors like soil type, soil depth, soil fertility, drainage, water table etc.

must be carefully examined. Well developed mango growing areas popular as traditional mango belt should be of first choice, however, developing areas, potential areas and demand driven areas should also be taken into consideration.It is presumed that traditional mango belts have all the basic requirements for mango cultivation and its trade whereas in newer areas, the selected site should be near to main road and market. Any mistake in selection of site can not be altered after planting while modifications in other factors are possible.

For successful mango production, some undermentioned special directions need to be followed properly.

Select Suitable Soil Type

Soil plays a crucial role in nutrient uptake phenomenon of a mango plant. It is supposed to create a buffer of nutrients which support the plant growth and fruit development. Mango is not too particular to soil type provided it has good drainage.

A good orchard soil should have:

- High cation exchange capacity.
- Good buffering capacity and a stable pH.
- Low electric conductivity (CEC).
- Low and stable C/N ratio.
- High moisture-holding capacity, to store available water and supply it when needed by the trees.
- Low bulk density and good aeration.
- High hydraulic conductivity for good drainage.
- Good anchorage and stability for the plant.
- Highly porous with minimum variation in volume by shrinkage when it dries.

Earlier, it was thought that only high fertile soils are better for commercial production but, now it has been proved that the poor and shallow soils are equally good for mango. On these soils, the trees remain smaller and more manageable provided the soil moisture management tools are adopted properly. On the other hand, rich deep loam soils contribute to luxuriant growth, resulting into less flowering and fruiting and showing erratic or irregular pattern of crop. Presently, mango is grown on lateritic, alluvial, kankar and other types of soil, however, medium and well drained soils have been found more conducive. Poorly textured soils with fragmented stones organals have been found to produce good mango yield in eastern, central and west coastal parts of the country under proper management.On shallow soils of hill slopes, mango trees grow to a large size but the yields are not satisfactory. On the lateritic soils, the trees are smaller and on sandy loams trees are of medium height. The red soils and red lateritic soils are considered to be the best soils for mango. For development of bright reddish tinge on solders and fruit surface with proper sugar: acid blend, the red or lateratic soils containing 5 to 10 per cent lime and sufficient quantities of peroxide of iron have been found suitable.

Soil Depth Permits Roots to Venture for Nutrient and Water

Mango has very deep and strong root system thus, soils for mango should be quite deep for easy penetration and spread of the root system. Soils with a depth of 1.2 m or more are ideal for mango crop. The deep and well drained soils with no impervious layers, allow good depth and distribution of root system producing trees of standard size, heavy yields and long life. More the depth of soil better will be plant performance.For high density planting where dwarf stature is one of the criteria, 60-100 cm depth is sufficient.

Structure of the soil in different horizons should be without impenetrable pans such as Kankar, rock, and compact structure like lateratic blocks and murram layer. The presence of a substratum of loose gravel or murrum helps in providing good drainage near feeder roots. Under such circumstances, the top 1 m soil should essentially permit the penetration and proliferation of plant roots.

Soil Inherent Nutritional Status

A good crop of mango (15 t/ha), removes 100 kg N, 25 kg P and 100 kg K.Thus, if not supported, the yield level and soil nutrient reserve are continuously reduced. Most of the mango-growing soils in India have a low soluble salt content, ranging from 0.04 to 0.05 per cent, total P from 0.06 to 0.0605 per cent and available K from 0.008 to 0.0087 per cent. Sandy soils are poor in organic matter content, nutrients and water holding capacity. While growing mango in such soils, texture of pit soil and plant basin should be improved by initial and continuous application of organic manures like farm yard manure or compost.

Avoid Abnormal Soil pH

Alkaline, poorly drained soils with rocky substratum and saline soils are generally not suitable for mango cultivation.Such soils need to be reclaimed by leaching out of salts using good quality water, replacing the harmful sodium from the soil with calcium or by establishing effective drainage system to avoid salt build up. However, these are costly and the expenditure is likely to recur as a continuous threat of salinity faced year after year. Mango is rated as moderately salt tolerant crop and therefore, its cultivation should be restricted in soils having very low (0.04-0.05 per cent) content of water soluble salts. Highly calcarious soils having large quantities of lime are considered poor for mango cultivation. Such soils are rich in sodium and become impenetrable to water. The presence of small amounts of kankar in neutral or slightly alkaline soils up to pH of 7.5 may not harm the tree; however, a pH range of 5.5 to 7.0 is ideal for mango growing. At pH values lower or higher than 6 to 7.5, the trees may suffer element deficiencies, especially phosphate and potassium.The soil pH may be got tested in local laboratories. The soil sample may be collected as per method given in the appendix. The abnormal pH may be rectified before taking up the plantation.

High Water Table is Harmful

A constant water table is ideal for good growth and development of mango crop. The water table should always be below a depth of 1.80 to 2.40 m because under high water table conditions, long time submerged feeder roots produce chloratic patches on leaves. Under this situation, the plant may not die but remain unhealthy and growth of the plant is adversely affected. During reproductive phase, high water table affects flowering and fruit set because of disturbance in tree physiology. However, lands with poor drainage and high water tables may also be utilized for mango cultivation provided that special techniques are followed in land preparations. Under such conditions trees may be planted on raised beds, about 1.5- 2 meters from the base level which will create a well drained space for trees to have root growth. In this case however, a good irrigation system must also be designed to water the trees as simple methods such as flood, basin techniques may be difficult to follow. Water logged situations lead to poor aeration of roots which may turn black, rot and ultimately die. It is a common experience with the trees which are subjected to floods or low lying areas having poor drainage generally die quickly. The land should be slightly elevated with little slope towards the drain. The drained and basin shaped areas should be avoided because such areas collect more rain water and become water logged. Elevated upland locations are more ideal for mango cultivation.

Suitable Climate Favours Mango Plant

Mango can be grown in both tropical and sub-tropical climate. For growing mango on a commercial and profitable scale, the temperature and rainfall have to be well within a clearly defined range. In addition, altitude, diurnal variation in temperature, rainfall occurrence, wind velocity and vapour pressure also influence growth and production of mango. Mango thrives well under humid and dry conditions. It requires good rainfall during its growing season i.e. June to October and rainless dry weather from November onwards. Rainy or cloudy weather during flowering favours the incidence of powdery mildew and leafhoppers. Some special considerations for climate suitablity of mango have been mentioned as "special hints" which may help in decision making.

High Altitude Leads to Low Mango Productivity

The plant grows luxuriantly and yields from sea level to 1000 m above mean sea level. However, for commercial cultivation, 600 m altitude is ideal. As the altitude of the place increases over and above 1000 m from MSL, the growth and productivity of the crop declines. The altitude and latitude has a definite role on the time of mango flowering. Flowering is delayed by 4-5 days at every 12m increase in altitude from 600m MSL at a particular place. Similarly, for each degree latitude, south or north of the tropics, flowering is delayed by four days.

Harsh Temperature Regime Affects Tree Productivity

Temperature governs the flowering time intensity and duration, sex ratio, fruit set, fruit growth and development and ripening time of mango fruits. In India, mango flowering takes place between January to March in northern parts due to onset of high temperature early in the season. Fruit ripening in these areas are also earlier than in north-western parts. Under moderate temperature conditions of Southern India, the flowering occurs in September to November. Fulfillment of required heat unit of a particular shoot and prevalence of minimum temperature difference between day and night and optimum humidity, some time leads to off-season bearing in mango. In general, the annual mean temperature for mango is around 23.90-26.70^{0}C. Mango trees grow and produce well in areas with very high temperatures (45 °C). However, when the maximum temperature exceeds 46 °C vegetative growth ceases, especially if it is accompanied by low humidity. For optimum growth and production, the average maximum temperature should be between 27 and 36 °C. When atmospheric temperature is high, and fruits are exposed to direct sunlight under air temperature of over and above 40.50 C, it develops spongy tissue in fruit. Abrupt high temperature coupled with low atmospheric humidity and high wind velocity of hot and dry winds cause severe splitting of mango fruit particularly in the light soils. Low temperatures particularly when the trees are in full bloom can cause the fruit to develop to approximately golf-ball size, turn yellow and then abort. Large numbers of these fruits result in a reduction in yield.

Rain Fall Pattern & Humidity Affect the Tree Productivity

Quantum and timing of rainfall play a crucial role in growing mango. In India, mango grows well at the places with 25-250 cm annual rainfall. However, in 75 cm and above annual rainfall areas, it can be grown as rainfed. Rainfall during flower bud differentiation (Nov-Dec.), flowering (Feb-March) and fruit ripening (May-June) and rains with hail storm during flowering and fruit development stage is harmful. Therefore, while selecting the site these climatic factors should be considered and such areas should be avoided. One of the pre-requisites for successful growing of mango is the presence of clear sunshine during the flowering and fruit set. Rain at flowering not only washes away the pollen, which adversely affects fruit set, but also encourages greater incidence of mango hoppers, mealy bugs and diseases like powdery mildew and anthracnose, which damage the crop severely. Cloudy weather with increased humidity in the atmosphere also encourages greater incidence of such pests and diseases. Bad weather also interferes with the activity of pollinating insects, thus adversely affect fruit set. In areas of excessive rainfall and high humidity at the time of fruit maturity and ripening, mango fruits are severely affected by fruit fly. High temperature, rainfall and humidity throughout the year, lead to continuous growth of shoots. Such shoots do not get requisite period for physiological maturity and thus, do not produce healthy panicle and fruits.High temperature and humidity in the coastal region of Kerala, Orissa and Andhra Pradesh lead to incidence of pests like stone weevil which restricts the export of fruits.

Frost Prone Areas are Harmful

Frost is common feature in northern and hilly areas. It affects mango plants adversely. The extent of damage by frost depends upon age of the tree, moisture content of the soil, condition of growth, actual timing of frost occurrence, severity and duration of the frost. Well grown trees with mature wood and in dormant growth phase are less affected than young trees which are immature under active growth phase. Level of soil moisture affects the extent of frost injury. Moist soil or irrigated soil raises the soil temperature and provides protection against mild frosts therefore; trees of such growth and age growing under dry soils conditions are severely damaged by frost than to those growing in wet land. The young mango plants are affected more severely by the early frost as they are not fully dormant and also by the late frosts owing to reinitiating growth and tenderness of the tissues. In old plants the frost injury is not very common, however, it is exhibited through bark splitting particularly when the wood is exposed (in case of rejuvenated plants). Oozing of gum from bark, death of new shoots, charred appearance of leaves and burning appearance of developing fruit buds are the main symptoms of frost injury. The temperature below1.1^{0}C in the night, bright sunshine during day and sharp drop in evening temperature with no wind are the pre-condition of frost. In bearing mango plants, the differentiating and exposed flower/fruit buds in bud-break and bud-burst stages are killed due to frost. They turn blackish and drop down.

Hail Storms Destroy the Crop and Plant

The hailstorms which often occur sporadically are main disasters during flowering and fruiting.The ice pallets destroy the panicle, developing flowers and small fruits.The damage caused to larger fruits is by physical hitting on the fruit surface which leads to rupture of tissue and such areas get discoloured and start rotting. Secondary infection by fungus starts some time after the physical damage. Affected fruits do not ripe and are unfit for consumption. Such fruits are not even fit for pickling. There is no method to protect hail storm damage.

Orchards in High Wind and Cyclone Prone Areas Need Special Attention

Hot/cool and strong winds are harmful to mango crop. In fact, high velocity of wind affects mango trees in several ways. High wind velocity leads to quick evaporation of water from the soil and thus reduce moisture availability which is very much necessary for ideal growth and development of the plant. Mango trees normally assume a dome shaped canopy and are comparatively less prone to damage of limbs and branches due to wind however, owing to weak attachment of peduncle, a large amount of fruit drop takes place due to wind. Strong winds (more than 120 km/hour), gales or storms blow away the branches, leaves or panicles and sometimes fully grown trees also get toppled. It is more so in the areas adjacent to coast or open areas that are not fully protected with windbreaks. Dwarf mango varieties like Amrapali, Rumani and Manjeera are less prone to wind damage compared to tall and huge trees like Langra. High velocity of wind may uproot the whole tree. This problem is commonly noticed in areas frequently affected with heavy cyclone especially in coastal areas of Orissa, Andhra Pradesh and Tamil Nadu. Fruit drop at various stages of its growth can be noticed due to high velocity of wind. At later stages of fruit development, fruit drop becomes severe due to heavy weight of the fruit which can be solved by the raising of thick and strong windbreaks around the mango orchard.

Special Hints Pertaining to Mango Growing Conditions

- Elevation is notmore than 600 msl. Fruit maturity is delayed at higher elevations.
- Rainfall
  - Minimum of 2-3 months dry period (less than 50 mm rainfall per month) at fruiting until harvest.
  - Less than 200 mm rainfall and/or 15 rainy days per month at flower induction.
- Relative Humidity is lower than 80% at flowering until harvest.
- Soil Moisture
  - Moist (<1> 10 atm.) at fruit maturation until harvest (110-130 days after flower induction).
- Near source of water for irrigation during dry months.
- Soil pH
  - pH range of 5.5-7.5.
  - Apply lime in soils with lower pH and use ammonium sulfate as N source at pH >7.

- Soil depth, texture and fertility
  - Deep (over 1 m) and well drained clay loam or lighter soil texture.
- Soil analysis: Organic matter (>2%), P (>15 ppm), K (>100 ppm), Ca (>2000 ppm), Mg (>200 ppm), Zinc (2-15 ppm), Boron (> 1-2 ppm), Sulphur (> 12 ppm) but no adverse reactions to low levels of sulphur is seen, Sodium (< 1 ppm),Chloride (< 250 ppm), Copper (0.3 - 10 ppm), Manganese (4-50 ppm), Iron (4-100 ppm), Conductivity (< 0.2).
- Full sunlight, long sunny days (high solar radiation) at flowering and fruiting.
- Topography is flat to rolling preferably below 20° slope.
- Mean temperature range of 24°C to 27°C.
- Preferably free from typhoon or cyclone; protected from dry and hot air.

Selection of Variety

Establishment of mango orchard is a long-term investment because mango trees continue to produce fruit for more than 50 years. The productivity of mango basically depends on variety and climate. The climate, sometimes leads to greater fluctuation in yield of a mango orchard but by and large it is the variety which governs the potential yield of the orchard. Therefore, a right variety should be selected for plantation at a particular place.Almost all our mango varieties have been selected from the naturally occurring superior chance seedlings based on their horticultural traits and fruit quality. Although, all the mango varieties have a wide range of adaptability, yield and quality performance is area, region and location specific.

Criteria for Selection of Mango Varieties

None of the existing commercial varieties of mango can be called as an ideal, as these lack one desirable character or the other. In general, the varietal selection is frequently made on the basis of yield potential, disease resistance, suitability for local conditions and availability of plant material in a near by place and often without sufficient consideration of market preferences and post-harvest behaviour particularly the keeping quality. In order to meet the challenge of the present-day requirements, the following criteria should be considered while selecting a mango variety.

i. Trees should have dwarf stature so that these can be planted at closer distance to harvest for optimum crop per unit area.

ii. The variety should have medium size fruits (200-250g per fruit). Golden apricot colour on ripening is desirable.

iii. Red blush on the cheeks and fibreless flesh, uniform consistency of the texture, proper sugar-acid blend and pleasing aroma are desirable characters for the varieties grown for export.

iv. High pulp: stone ratio and firm flesh are the highly desirable characters.

v. The variety should have regular bearing capacity to ensure consistency in return.

vi. The variety should be highly tolerant to various fungal and bacterial diseases, insect pests and physiological disorders like mango malformation, spongy tissue, black tip, internal break down. However, any such variety with multiple resistance is lacking.

vii. The variety should have market demand. At a particular location, if a suitable variety cannot be grown with satisfactory yield, it is better not to go for that. When the market is not very much specific and more than one variety meet market requirements, then the variety should be chosen which has higher yield and the most favourable post-harvest characteristics.

viii. Stable pleasant flavour should be characteristic of a variety particularly when it undergoes processing.

At present none of the commercial varieties of mango has all these characteristics combined into one.As a matter of fact, all these characters may not be available in a single variety or if at all it is available, the variety may not perform well in all the mango growing areas. Therefore, it is essential to choose a wider adaptable variety, because the success of any mango plantation depends on tree health and its management. Adaptable variety and its compatible environment is the foundation on which tree management systems are built up. When the right variety is in place, tree management would facilitate enhancing the tree productivity and orchard profitability. However, when incompatible varieties are grown in an unmatched environment any tree management system, would hardly make any significant improvements in fruit yield. If growers at a particular place are interested for new varieties, especially the exotic which have good market demands, it is better to first analyze the growing environment of those varieties with that of the local environment where the orchard is going to be established.

Some Mango Varieties are Region Specific to An Extent

There are more than thousand mango varieties in India. These varieties are region specific to some extent. For instance, variety 'Langra' performs very well under in north India but flowers and fruits very sparsly under South Indian condition. Similarly, the South Indian varieties do flower and fruit under north Indian conditions but some of their characteristics do not remain same. For example, Neelum which is a regular bearing South Indian variety with bigger fruit size although tends to bear every year, the fruit size is markedly reduced and ripening is delayed in north India . Likewise, 'Rumani' variety of South India produce less female flowers, resulting in sparse fruiting under North Indian conditions. Alphonso of Ratnagiri and Kesar of Gujarat do not produce same colour and quality under North and Eastern Indian conditions as in the coastal region. Bearing potential of Bamisan of Andhra Pradesh is not fully depicted in other parts of the country. The variety produces mixed flush in North and Eastern India and quality of fruit is also reduced. Thus, commercial varieties of mango, although having a wide range of adaptability, are specific to different regions of the country for commercial cultivation point of view. Nearly 30 varieties are grown on commercial scale in different states. These are presented in following table-4.

Table 4: Important mango varieties of different states of India

States	Varieties
Andhra Pradesh	Banganpalli, Cherukurasam, Himayuddin, Suvarnarekha, Allumpur, Baneshan, Neelum, Totapuri.
Bihar	Bombay Green, Langra, Fazli, Himsagar, Kishenbhog, Malda, Sukul, Bathua, Gulab Khas, Zardalu, Chausa, Dashehari,.
Goa	Fernandin, Mankurad, Alphonso.

Contd...

Table 4. Contd...

Gujarat	Alphonso, Kesar, Rajapuri, Vanraj, Jamadar, Totapuri, Neelum, Dashehari, Langra.
Haryana	Dashehari, Langra, Bombay Green, Sarauli, Chausa, Fazli.
Karnataka	Alphonso, Mulgoa, Neelum, Pairi, Baganapalli, Totapuri.
Kerala	Mundappa, Olour, Pairi.
Madhya Pradesh	Alphonso, Bombay Green, Langra, Sunderja, Dashehari, Fazli, Neelum, Amrapalli, Mallika.
Maharashtra	Alphonso, Kesar, Mankurad, Mulgoa, Pairi, Rajapuri, Gulabi, Vanraj.
Orissa	Baneshan, Gulab Khas, Langra, Neelum, Suvarnarekha, Lat Sundary, Banganpalli, Cherukrasam, Amrapalli, Mallika, Totapuri.
Punjab	Dashehari, Langra, Chausa, Malda.
Tamil Nadu	Banganpalli, Neelum, Rumani,Mulgoa, Alphonso, Totapuri.
Uttar Pradesh	Bombay Green, Dashehari, Fajli, Langra, Safeda, Samarbehisht, Chausa.
West Bengal	Bombay Green, Himsagar, Kishen Bhog, Langra, Malda, Fazli, Gulab Khas, Amrapalli, Mallika.
Himachal Pradesh	Chausa, Dashehari, Langra.
Jharkhand	Jardalu, Amrapalli, Mallika, Bombay Green, Langra, Himsagar, Chausa, Gulab Khas.

Broadly, region wise prevalence of mango varieties is as follows:

- *North India :* Dashehari , Chousa , Langra, Mallika, Amrapali
- *South India :* Totapuri , Neelum, Banganpalli
- *West India :* Alphonso, Kesar, Ratna, Rajapuri
- *East India :* Langra, Himsagar, Fazali, Malda, Dasheri, Alphonso, Totapuri, Kesar

Maturity Period of Different Varieties Vary

In India, the mango varieties mature from mid of March to mid of September except two cases of Baramasi or off-season clones which although do not have desirable pulp quality and yield potential, matures by October-November. It has been established that the regions experiencing uniform temperature pattern throughout the year or in other words do not have clear cut winter month, produce mango crop early. Apart from the climate, the genetical characteristics of a variety decide the early, mid and late crop of mango. Based on time of maturity and ripening, the commercial mango varieties have been classified in early, mid and late groups (Table-5).

Table 5 : Classification of mango varieties on the basis of maturity

Maturity	Group
Early	Gulab Khas, Bombay Green, Himsagar, Kesar, Suvernarekha, Jardalu.
Mid-season	Alphonso, Mankurad, Totapuri, Vanraj, Banganapalli, Dashehari, Langra, Kishen Bhog, Mankurad, Mallika.
Late	Fazli, Fernandin, Mulgoa, Neelum, Chausa, Amrapali.

Select a few Among Many Varieties Available in the Country

Mango varieties and hybrids vary in their shape, size, colour, taste, regional suitability, yield, response to post harvest management practices, commercial use and pest and disease problems. For reference to growers the important features of some of the conventional varieties of mango in the country has been enumerated as follows:

1. *Alphonso* : Mid season leading commercial variety of Maharashtra state and one of the choicest varieties of the country. Also known as Badami, Gundu, Khader, Appas, Happus and Kagdi Happus in different regions. The fruit is medium in size, ovate oblique in shape and orange yellow in colour. The fruit quality is excellent and keeping quality is good. It has been found good for export as well as canning purpose. It has limited adaptability in the Western and South western parts of the country. The flesh develops spongy tissue and the plant shows biennial habit.

2. *Totapuri* : Mid season commercial variety of south India. Also known as Bangalora and Collector at different places. Heavy

yielder and regular bearing variety. The fruit size is medium to large and shape is oblong with necked base. The mature fruit colour is golden yellow. Fruit quality is poor but the keeping quality is very good. It is widely used for processing and susceptible to bacterial spot.

3. *Banganpalli* : It is a commercial variety of Andhra Pradesh and Tamil Nadu and has better choice among the early arrivals in the market. It is a mid season variety but fruits can be pushed in the market little early. Also known as Chapta, Safeda, Baneshan and Chaptai. Fruit is large in size and obliquely oval in shape. The colour of the fruit is golden yellow. Fruit quality and keeping quality are good. It is good for canning and pulp processing.The plant show weak tree form and biennial habit.
4. *Bombay Green* : Bombay Green is a commercial early season variety North India. It is also called Malda, Bombai, Sabuja at different places in Northern India. Fruit size is medium, shape ovate oblique and fruit colour is spinach green which some times turn yellow particularly in fruits exposed to sun light. The fruits have prominent lenticels. Fruit quality is good and keeping quality is medium. Its taste and flavour are excellent and serves as a good blend for mango nectar. It is a shy bearing variety and have strong tendency towards alternate bearing . It is highly susceptible to vegetative and floral malformation.
5. *Dashehari* : It is a leading commercial variety of north India and one of the best varieties of our country. It is a mid season variety and is mainly used for table purpose. The fruit size is medium, shape is oblong to oblong oblique and fruit colour is yellow. Fruit quality is excellent and keeping quality is good. Its commercial bearing starts from 4th year of planting and the plants have high pruning response and thus, the variety can be planted in closer planting with proper maintenance. Normally, the plant shows biennial bearing habit and high susceptibility to malformation.
6. *Fajli* : This variety is commonly grown in the states of Uttar Pradesh, Bihar and West Bengal. This is a late season variety. Fruit is very large, obliquely oval in shape. Fruit colour is light chrome and keeping quality is medium. The fruits are abnormally big in size, inferior in quality and the plant is biennial in habit.
7. *Fernandin* : Fernnadin is one of the oldest varieties of Bombay. It is a late season variety mostly used for table purpose. Fruit size is

medium to large, fruit shape is oval to obliquely oval and fruit colour is yellow with a red blush on shoulders. Fruit quality and keeping quality are medium.

8. *Himsagar* : Himsagar is indigenous to Bengal although it produces good crop in Bihar, Jharkhand, Uttar Pradesh and Orissa. This is one of the choicest varieties of Bengal and has gained extensive popularity. Fruit is of medium size, ovate to ovate oblique. It is an early variety. Fruit colour is greenish yellow at ripening. Fruits are very sweet and pulp has very less fiber with better shelf life. Fruits are susceptible to black tip disorder under acidic soil conditions and tree shows biennial bearing habit.

9. *Kesar* : Kesar is a leading variety of Gujarat with a red blush on the shoulders. It is an early variety. This variety has export potential. Fruit size is medium, shape oblong and keeping quality is good. Kesar is widely adaptable variety of mango but the desired colour and quality of fruit is not produced in northern and eastern parts of the country. A clone of Kesar identified by Reliance at Jamnagar in 2003-04, named as Ril Kesar .The fruit size is around 400 g . All fruit qualities are same as Kesar . Similar clone identified by a local farmer and named as Jumbo Kesar.

10. *Kishen Bhog* : This variety is indigenous to Murshidabad in West Bengal. It is a mid season variety. Fruit size is medium, fruit shape is roundish oblique and fruit colour is yellow. Fruit quality and keeping quality are good. Plants are susceptible to gummosis in acidic soil of Eastern India.

11. *Langra* : This variety is indigenous to Varanasi area of Uttar Pradesh. It is extensively grown in northern India. It is also known as Maldha in Bihar and West Bengal. It is a mid season variety. Trees are vigorous and spreading. Fruit is of medium size, ovate shape and lettuce green colour. Fruit quality is good. Keeping quality is medium. Now it is becoming a popular variety as it has wide adaptability. Plant take more time to start bearing and poor fruiting is reported in young age due to excessive fruit drop. It is strictly biennial in habit.

12. *Mankurad* : Mid season commercial variety of Goa and in the neighboring Ratnagiri district of Maharashtra. Fruit is medium in size, ovate in shape and yellow in colour. Fruit quality is very good but the keeping quality is poor. The variety develops black spots on the skin in rainy season.

13. *Mulgoa* : Mulgoa is a commercial variety of southern India. It is quite popular among the lovers of mango owing to high quality of its fruit. It is a late season variety. Its is spread limited to Southern India. Fruit is large in size, roundish oblique in shape and yellow in colour. Fruit quality and its keeping quality is good.
14. *Neelum* : Late season commercial variety of Tamil Nadu and adjoining areas. It is an ideal variety for transporting to distant places owing to its high keeping quality. Fruit is medium in size, ovate oblique in shape and saffron yellow in colour. Fruit quality is good and keeping quality is very good. It is a regular bearing variety under South Indian condition.The plants of Neelum becomes dwarf under north Indian conditions and have inferior fruit quality.The fruits are susceptible to bacterial canker.
15. *Chausa* : It is commonly grown in northern parts of India. It is also known as Summer Behist Chausa. It is a late variety. Trees are quite vigorous and show apical dominance leading to erratic bearing. Fruit is large in size, ovate to oval oblique in shape and light yellow in colour. Fruit quality is good and keeping quality is medium. Unpredictable shy bearing particularly in young age, biennial in habit and very high incidence of floral malformation are some of the demerits of this variety.
16. *Suvarnarekha* : This is a commercial variety of Visakhapatnam district of Andhra Pradesh. This variety is also known as Sundari, Lal Sundari in Andhra Pradesh and Orissa. It is an early variety. Fruit is medium in size and ovate oblong in shape. Colour of the fruit is light cadmium with a blush of jasper red. Fruit quality is medium and keeping quality is good.
17. *Vanraj* : Vanraj is a highly prized variety of Vadodra district of Gujarat and fetches good returns. It is a mid season variety. Fruit is medium in size, ovate oblong in shape and colour is deep chrome with a blush of jasper red on the shoulders. Fruit quality is medium and keeing quality is good.
18. *Zardalu* : This variety is indigenous to Murshidabad in West Bengal and has great demand in Bihar, Jharkhand. Fruit size is medium, oblong to obliquely oblong and golden yellow in colour. Fruit quality is very good. Keeping quality is medium.
19. *Dashehari-51:* Developed at CISH, Lucknow.
20. *CHM-7:* Developed at CHES, Bhubaneswar. Selection from Gulabkhas, early maturity and good external colour.

Hybrid Varieties

1. *Amarapali* : It is dwarf, regular and cluster bearing hybrid which matures late. The fruits have good keeping quality. The variety responds to pruning and therefore, suitable for high density planting accommodating even 1600 plants in a hectare. It yields on an average 16 tonnes / hectare. It is susceptible to malformation and bears uneven size fruits.
2. *Mallika* : Its fruit is large in size, oblong elliptical in shape and cadmium yellow in colour. Fruit and keeping quality are moderate, high TSS. It is a mid season variety and reported to be a regular bearer. Fruits are susceptible to bacterial canker.
3. *Arka Aruna* : It is a dwarf, regular bearing, precocious hybrid. Fruits are large sized (500-700) with attractive skin colour. Pulp is fibreless, sweet to taste (20-22 Brix). Pulp percentage is 73 and the fruits are free from spongy tissue.
4. *Arka Puneet* : It is a regular and prolific bearing hybrid .Fruits are medium sized (220-250) with attractive skin colour, having red blush. Pulp is free from fibre, pulp percentage being 70 per cent. Fruits are sweet in taste (20-22^0 Brix) with good keeping quality and free from spongy tissue. It is a good variety for processing also.
5. *Arka Anmol* : It is a semi-vigorous plant type .It is also a regular bearing and free from spongy tissues. Fruits ripen to uniform yellow colour. Keeping quality of the fruit is very good and it is suitable for export. It has got excellent sugar and acid blend and fruits weigh on an average about 300 g. Pulp is orange in colour.
6. *Arka Neelkiran* : It is a regular bearering late season variety with medium sized fruits having attractive red blush and free from spongy tissue.
7. *Ratna* : Tree vigorous, fruits are medium sized, attractive in colour with excellent quality. It is a regular bearer and free from spongy tissue. Flesh is firm and fibreless, deep orange in colour with high TSS (19-21^0 Brix).
8. *Sindhu* : It is regular bearer, fruits medium sized, free from spongy tissue with high pulp to stone ratio and very thin and small stone.
9. *Au Rumani* : It is precocious, heavy and regular bearer with large fruits having yellow cadmium skin colour. Flesh is moderately firm.

10. *Manjeera* : It is a semi vigorous type, regular and prolific bearer hybrid. Fruits are medium in sized with light yellow coloured skin, firm and fibreless flesh and sweet to taste.
11. *Sundar Langra* : It is a regular bearing hybrid resembles Langra. The fruits have lighting over the solders.
12. *Alfazali* : It is a cross between Alphonso and Fazli and found superior to Fazli. No spongy tissue is found.
13. *Pusa Surya*: Developed at IARI, New Delhi.Apricot yellow with red blush, regular, heavy fruiting , late .
14. *Pusa Arunima*: Developed at IARI, New Delhi. Regular, heavy fruiting, fruits are of red colour and plants are semi vigorous.
15. *CISH M1* : Developed at CISH, Lucknow. Fruits are of yellow colour with red blush, firm and fibreless.
16. *CISH M2* : Developed at CISH, Lucknow. Fruits are dark yellow, firm fleshed and late in maturity.
17. *Nileshan* : Developed at ARS, Paria, Gujarat. Regular, heavy and late fruiting.
18. *Neelphonso*: Developed at ARS, Paria, Gujarat. Regular, moderate fruiting,smaller fruits.
19. *Nileshwari* : Developed at ARS, Paria, Gujarat. Dwarf, regular and late fruiting.
20. *Sonpari* : Developed at ARS, Paria, Gujarat. Regular, excellent quality and spongy tissue free.
21. *Neelgoa*: Developed at ARS, Ananatrajupeth, Andhra Pradesh. Regular and moderate fruiting, bigger fruits with red blush.

Sucking Type Varieties

Apart from table varieties, there are quite few sucking types characterized by juicy and soft flesh with fibres. These are not yet cultivated on commercial scale but grafted trees are grown in the orchards of big mango growers. Some such varieties are 'Raspooria', 'Mithwa', Sundar Shah', 'Mithwa Ghazipur', 'Taimuriya', 'Sharbati Begrain','Gilas', Nauras','Rasgola','Har-dil-aziz','Cherukurasam' and 'Peddarasam'. Fruits of all these varieties are small to medium but these varieties are mostly biennial in habit.

Coloured Varieties

Colourness adds to the desirable characteristics of a variety and enhances consumer appeal. However, almost none of the commercial varieties of mango in India has this component of red blush. The varietal wealth of mango in India is huge and a number of varieties are known to have brilliant red blush on their cheeks. Some well-known varieties having red colour are 'Gulab Khas', 'Sinduriya', 'Surekha Calcutta', 'Zafran', 'Husna-e-ara', 'Janardhan Pasand, 'Suvarnarekha, 'Lal Mulgoa' and 'Vanraj'.

❑❑❑

4

Learn to Produce and Use the Quality Planting Material

The seedling trees have long juvenile period and have more vigorous growth habit, which creates difficulty in taking plant protection measures and harvesting of their fruits. The fruits of seedling trees do not mature in one stroke and thereby affect their marketing. Mango is cross-pollinated crop and produces heterozygous seeds and therefore, propagation by seed leads to enormous variability in the progenies.Keeping these disadvantages of seedling trees in mind and to obtain uniformity in plant performance, a cultivar should be propagated through asexual methods of propagation. Knowledge of mango propagation is required for following purposes.

1. Development of *in-situ* orchard in an area where expert grafter is not available or even the grafter is hired for bulk grafting, need may arise in future to re-graft the failed grafts where rootstock is surviving and scion is dead. If farmer has knowledge, he can manage the the rootstock/orchard to make uniform plantation through grafting in emergency.

2. In many cases, the rootstock survives but scion is dead. This situation arises in many circumstances such as planting of immature grafts, planting at inappropriate time, faulty planting technique and poor post planting care. A farmer having knowledge of propagation can manage the failed grafts by his

own to maintain orchard uniformity especially when expert grafters are not available readily.

3. Many old orchard need rejuvenation where bulk as well as recurring grafting may be required. For this purpose grafter may not be available all the time however, a farmer having knowledge can manage the situation.
4. The cost of self propagated sapling is less than purchased one. Further, self propagated planting material ensures the genuineness and truthfulness of the variety. A farmer, if wish to reduce the cost on sapling, can prepare plants by himself provided that he has expertise.
5. Nursery business is a very lucrative aspect in mango production. However, a nursery operator should know the basic and practical aspects to be able to produce good quality planting materials as it is the key to economic viability of commercial nursery.

For the benefit of all above said partners, the basic concept of nursery is given hereunder. Important factors considered for establishing a nursery are agro-climatic conditions, soil types, soil pH, location, area, irrigation facilities, communication, market demand, availablity of mother plant and skilled persons etc.

Knowledge of the Nursery Act is Needed

Most of the states have now enacted Nursery Act. This act varies from state to state. Nurserymen should know about it. First nurseryman has to obtain license from DHO then he should procure mother plants from ICAR institutions/Agricultural University or Govt. nurseries. He has to label mother plants and grafts raised as these nurseries are inspected by Govt. Horticulture Officers. Further, plants are to be sold at prescribed rate and nurserymen have to submit periodic reports to Govt. Horticulture Officer. A nurseryman is authorized to produce only certain number of mango grafts based on number and age of the mother plant, he has maintained.

Selection of Site

The site selected for raising a nursery should preferably be located near marketing centres for the convenience of transportation of the plants with minimum or no damage. The site should be convenient enough for transportation of input materials also. It is important to have or develop a perennial source of water inside the nursery. If need arises, wind breaks of tall plants like neem, tamarind, aonla, seedling mango, etc. may be planted to provide necessary shade and protection.

Selection of Varieties

The varietal choice primarily depends on the market demand in nearby areas. For wider market coverage, the choice may be dependent on market studies in the desired areas.

Establishment of Nursery Structure/Materials and Training to Workers

For raising rootstock seedlings, some permanent or temporary structures for seed bed may be prepared. These beds should be 0.5 to 0.75 m high from ground level. The beds may be 0.75m to 1.00m in breadth and length may be as per the availability of land.The required structures and facilities include a nursery shed (storage room or potting area), a compost pit, and a water system which involves a tube well, and a pump, pressure tank, and fittings. The workshed with thatch roofs and locally available materials like bamboo, wood, etc. may be constructed. For potting during rainy season, the soil and compost should be stored in hot or summer months as in rainy season, collection of dried soil and manure is very difficult and without these, the seedlings cannot be raised.

The propagated plants are planted in nursery beds for better growth or hardening of the plants. In general, this type of nursery bed is prepared under partial shade. For better sprouting, polyhouse of expanded metal and polythene roof supported by local materials like bamboo, wood and planks, may be constructed. The grafted plants in polythene are also kept in net house made with bamboo / iron bar and shading net. A store-cum-office of 6.0m x 4.5 m constructed with locally available materials may also be needed. Fencing is needed to protect plants from animals.

Equipments and tools such as pruning secateurs, garden rakes, khurpa (pvc and wooden grip), bill hook, garden saw, garden sword, sprinklers with heavy duty, stand and garden tool kit, pickaxe, watering can, crowbar, mugs, trowels, large funnel, wheel barrows, shovels, pruning shears, grafting knives and knapsack sprayers may be needed in nursery. The other consumables materials include black plastic bags, urea, insecticide, fungicide, river sand, bamboo poles, plastic bags, garden soil, compost, polyfeeds and scions.

Labuor force is needed for land clearing and leveling, hauling, dehusking, and sowing seeds, plastic bagging of seedlings, watering, applying fertilizer, weeding, spraying, preparation and collection of scions, removing from plastic wrap and covering living grafts. Hence, they should have some basic idea of nursery activities. A well trained grafter is also needed.

Development of Mango Scion Block

An essential need in propagation of mango is to have a reliable and accessible source of quality mango scions. This can be done by establishing a mango scion block in one area of the nursery and maintaining it properly. The mother plants must be true- to the- type and true -to the- variety. The plants should be properly labeled. The mother plants may be maintained properly for their vigorous growth; otherwise number of propagated plants will get reduced. The mother plants may be planted as commercial orcharding, however, high density planting is more suitable for mother plants.

Preparation of Rootstock

Rootstocks affect plant growth and tree performance. Besides, these are helpful in growing plants under adverse soil and climatic conditions. Disease and pest resistance obtained through specific rootstocks may be of special importance under certain situations. Recently, however, rootstocks have especially been recommended for inducing dwarfness in fruit trees. Rootstocks must be raised first to produce grafted mango plants. Generally mango stones are collected from local trees, market places etc. irrespective of the cultivar during the mango season. Bolder stones give seedlings with more vigour irrespective of whether these are collected from grafted or seedling trees. If extracted and kept for a longer period, stones lose their germination capacity completely. The fresh stones collected from processing units give high germination while stones embedded in peels or in sun for long period or dried stones are poor in germination. Stones sown within one month of extraction germinate about 80 per cent and none after that. Stones when stored in the open, lose viability completely after 70 days, whereas those buried under sand retain 10-15% viability after 90 days. Similarly, the seeds mixed with charcoal and sealed in tins or earthen pots retain their viability only for 80 days at room temperature.

Stones are sown in beds immediately after extracting them from the ripe fruits. Extracted seeds may have mango weevils inside. When such seeds are planted, whole seed may be destroyed. Therefore, dip the seeds in an insecticide solution such as endosulfan 0.07% before planting for a few minutes. Use of fungicides such as 'Captan' at this time is helpful to protect the seeds from fungal infections in the nursery. Seeds treated with insecticide and fungicide solution must be planted in a sandy seed bed. For the seed bed, 15-20 cm thick layer of sand may be used. The stones are placed without leaving any space among them keeping upright plumule so that straight tap root and stem is produced. After placing the stones, the moist leaf mould is placed over them. The seeds germinate within 15-20 days after sowing. After seeding, the sand bed is kept moist at all times until germination. Regular irrigation and weeding is practiced in nursery. The rootstock should not be grown in the shade as shades provide a microclimate which favors the growth of pathogen. Seedlings that grow without shades are hardy and free from fungal diseases. Seedlings do not need too much water. At high moisture levels, seedling may be infected with fungal diseases. Spray of chemicals may be done to protect young seedlings especially from leaf cutting weevil and diseases. If seed bed is containing materials such as coir dust, care must be taken to maintain the appropriate moisture level and management of termites.

Planting in Polyethylene Pots or Planting a Secondary Nursery

When the leaves colour changes from coppery red to green, the seedlings are shifted to permanent nursery beds or transplanted in polyethylene bags to check the root growth.Tetrabags made out of 200 gauge polyethylene to a dimension of 20 cm diameter and 30 cm height may be used. Potting mixture is prepared by mixing equal parts of topsoil, sand and well composted cow dung or leaf litter together. Use of one part of coir dust to four parts of above mixture is helpful to improve the water holding capacity of the potting media. For improving the growth of seedlings, light doses of inorganic fertilizers, viz., ammonium sulphate and calcium ammonium nitrate, at the rate of 22 to 27 kg may be given. Polyethylene containers are filled with the above mixture as to leave 2-3 cm from the top of the container. This is necessary to prevent overflow of water when pots are irrigated.

If seedlings are not directly transplanted in bags, it must be transferred to a secondary nursery. To make a secondary nursery, beds of about 20 cm in height are prepared and filled with a mixture of top soil and well composted cow dung. Drain is provided between two beds to remove the excess water. Double row system of planting should be adopted in the secondary nursery. In a single bed, two double rows are planted (four rows of seedlings). The seedlings should also be protected from frost by thatching. Plants in secondary nursery or polyethylene pots become ready for grafting 7-8 months after transplanting. The stem of the stock plant must attain pencil thickness before grafting.

When stock plants are transplanted into polyethylene bags directly and grafted, they become ready for planting in 7-8 months. However, when stock plants are transferred to a secondary nursery and grafted, it takes about 16 months to get it ready for field planting. Thus, the cost of production of such plants increases invariably. Field survival of such plants is also affected due to damage of root at the time of lifting from the secondary beds.

Selection of Scion

The mother trees which provide scion material are as equally important as the rootstocks. A scion is taken only from mother trees having true to type characteristics, good fruit quality and bears regularly. Scion wood must be collected from branches bearing fruit. Use of water shoots and mutated branch should invariably be avoided while taking scion material. For grafting, shoots with mature leaves around apical bud is the most suitable. Before grafting, the shoots must be conditioned. One week before grafting, all the leaves in the selected scion should be removed leaving only petiol. Then the defoliated scion stick is cut to about 10-15 cm length.

Propagation Techniques

Inarching : Seedlings are raised by using different methods. After about one year, they become ready for inarching. Healthy seedlings are transferred either to pots or taken out (with ball of earth around the roots tied with grass all round to check the evaporation of water and to keep the roots intact) for inarching operation. The scion shoot should be healthy with well-developed foliage. These seedling root stocks are taken to the scion tree and kept or tied up conveniently to facilitate the process of inarching. Sometimes the scion parent is headed back to produce a large number of conveniently available branches for grafting at ground level. A strip of bark (6.0-7.5 cm long) with a small layer of wood attached to it is removed from the potted seedling at a height of 22 cm from soil surface. A similar strip is then removed from the scion shoot selected for inarching. Care is taken to ensure that both the cut surfaces on the seedling and the scion shoot fit together perfectly, leaving no gap. In this position, they are tied firmly with banana fibre or polythene strip. When union is developed the lower part of scion material and upper part of rootstock are cut. Once the plant is ready, it may be shifted for further use.

Veneer grafting : This method is easier, more economical, gives a high percentage of success and is ideal for establishing *in situ* orchards. It is also known as detached method. In this method 1-2-year-old (50 mm diameter) rootstock is used. The scion of similar thickness having 3 to 4 months age, preferably a terminal non-flowering shoot is also selected. The selected scions are defoliated on the mother plant about 7 to 10 days prior to detaching. This method can be adopted from March to September, preferably in June-July. A slanting downward and inward cut, 2.5 to 4.0 cm long, is made on rootstock 15 cm above the crown portion of stock. At the base of cut, another shorter cut is given to intersect the first, so as to remove the piece of wood and bark. The base of scion is given a long cut along one side and a very short cut on the opposite side so as to match the cuts on the stock. The scion is inserted into the stock so that the cambium layers match on the longer side. The graft union is tied lightly with transparent polyethylene (100 gauge thickness) strip. When scion shoot starts growing and produces the vigorous sprout, the rootstock just above the graft union is cut back. Then care and maintenance of graft is done in nursery. The percentage of success is about 96 when done in June and July.

Side-grafting : It is commonly practiced in coastal regions. It is exactly similar to that of veneer grafting except that in veneer grafting the vertical flap of the root stock bark is completely removed whereas, in side grafting this flap is retained and tied over scion. Another difference between two methods is that only one side of the scion is sliced away in sloping manner in veneer grafting, whereas in side grafting the scion is sliced on both sides of the lower portion in the form of wedge.

Epicotyl/stone grafting : Germinating seeds of about 4-8 days old are used as rootstocks. The scions are prepared by prior defoliation of shoots of comparative thickness. Splice and wedge methods are used for grafting. For splice grafting the epicotyl is cut slantingly for 2-3 cm length and the lower portion of scion is also cut to match it. The cut surfaces of both, the stock and scion are tied together with polythene so that the cambium of each other comes in close contact. In wedge grafting, a vertical cut 4-6 cm long is given into the beheaded epicotyl, to receive the wedge-shaped scion. This is then tied with polythene film. The grafts prepared by these methods are planted immediately in pots and watered. Grafting is done in rainy season when there is high humidity in the atmosphere. The scion sprouts within a month of operation. The percentage of success in splice and wedge methods is low.

Soft-wood grafting : When 1-year-old seedlings raised for rootstock commence putting on new growth and the leaves turn bronze, these are ready for soft-wood grafting. Shoots of 3-4 months, which have prominent apical bud, are taken as the scion material. Leaf lamila from such shoots is removed about a week before detaching them from the parent tree. At the time of removal of these shoots, the apical bud should remain intact. The top of the new growth of the stock is cut and the scion is fitted by wedge grafting. The union is tied with 100 gauge polythene strip. If the selection of rootstock and scion is proper, success is 100 per cent when the grafting is done during July-August under north Indian conditions. This method is utilized for establishing in *situ* mango orchard in gravelly soils.

Cleft or wedge grafting : Primarily this technique is adopted for rejuvenation of old orchards by top working the trees. The rootstock, which is not fit for other grafting technique due to more thickness can

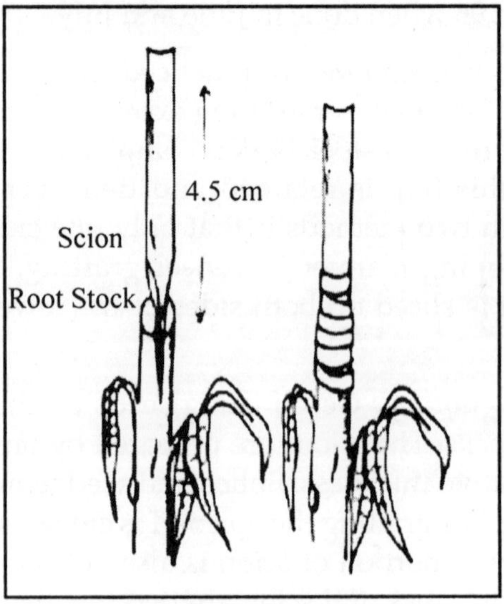

Fig. 2 : Cleft grafting

be used in this method for grafting. The stock is beheaded from any desired height and cut is made at right angle to the main axis of the branch. A vertical split 5.0 to 7.5 cm down the centre of the cut stub is made with the help of a sharp knife. The lower end of scion shoot is given two 5.0 to 7.5 cm long slanting smooth cuts on either side. The scion is inserted into split of rootstock, and tied with polyethylene strip.

Budding : This is the most economical method but uniform good success has not been obtained in different types of agroclimatic regions. The budding technique is highly suitable to those areas where general atmospheric humidity remains high. The most commonly employed methods are patch, shield and forkert budding. The rootstock seedlings are raised in similar fashion as done in case of grafting. The best-suited time of budding in North India is February to June while in South India, August to September. The bud wood/scion should be of last season growth and shoot should be round in shape having grey colour. The middle bud is considered best for good success.

Shield budding : This is one of the commonest methods used for propagation of mango in other countries. The term shield bud is derived from the shield like shape of the bud. Seedlings which have attained about one year age and pencil thickness in diameter are given a transverse cut about 15 to 20 cm above the ground level. The cut is kept just deep enough to cut the bark and about one-third distance around the stock. A perpendicular cut starting from the middle of transverse cut and about 2.5 to 4.0 cm long is made. Then bark is loosened on either side of the cut, without tearing the bark. To remove the shield of bud, a cut about 2 cm. and a horizontal cut at the required portion separates the bud shield which carries a thick piece of wood. Now shield piece containing a bud is inserted in stock plant. The bud is inserted in between the two flaps of bark without tearing the bark. Then the bud union is wrapped lightly with polyethylene strip. During tying the bud portion remains open. If the budding is successful, the bud piece unites with rootstock in two to three weeks time. The union is completed after about 30 days. The stock is cut back about 2.5 cm above the union. After a season's growth in nursery, plants are shifted to some other place before planting to their permanent location.

Forkert budding : In this method a transverse cut of 8 mm width is given on root stock at 25 cm above the ground level and two vertical cuts of about 2.5 to 4.0 cm on both sides of transverse cuts. The bark is removed carefully along these cuts. The bud is prepared in similar manner as in case of shield bud. The size remains equal to the cut on the stock. The bud is placed into the exposed portion of stock and the bud is covered by the flap, then budded portion is wrapped with polyethylene strip. After a month when union takes place, the wrapping material is removed and the flap is cut off. When the bud shows symptoms of growth, the rootstock portion is given half cut above the union leaving some leaves attached to the stock such operation stimulates the growth of bud. After attaining new growth from bud, the remaining portion of rootstock is cut from 2.5 cm above the bud union. Now plant is allowed to remain in nursery for one season before shifting to permanent location.

Patch budding : This method of mango propagation is common in Florida. A rectangular patch of bark of 2.5 cm long and 1.0 to 1.5 cm wide is removed from rootstock from 15-20 cm above the soil surface. A bud shield of similar size or slightly smaller is taken from the scion shoot and placed over the exposed portion of rootstock. Then tying is done with polyethylene strip. To get success in patch budding, it is essential that stock as well as scion should slip easily.

Manuring, Watering, Drainage and Plant Protection

Manuring is to be done very carefully. Vigorous growth of plant is always attractive to the buyer. Again, heavy manuring is not beneficial for storage of plants. Like manuring, watering is also important. Watering is done according to need of the plant. The nursery should have a water source of its own. Digging a well (12 m deep x 3 m dia) and installation of a 2.0 HP pump set with accessories are considered. Sprinkler system of irrigation is not advisable at the beginning.

For sufficient vegetative and reproductive growth of plants, good drainage system must be developed in between the beds and around the nursery. Adequately gentle slope in the pot bed surface is also desirable. It is extremely important to ensure that water logging does not occur in and around the pots and beds.

Keen observation on attack of different pests and diseases is required. If the mother plants are infected, the propagated plants will be infected also. Necessary control measures should be taken immediately on observation.

❑❑❑

5

Adopt Standard Orchard Establishment Practices

The success of a mango orchard mainly depends on the initial establishment practices.A well established orchard itself avoids many problems related to plant growth and performance.Therefore, one can prevent many risks through appropriate orchard establishment practices.This include:

Land Preparation and Appropriate Layout

Poor land preparation and inappropriate lay out posses recurring problem. Therefore, selected fields should be deeply ploughed and then harrowed to root out the perennial weeds and heavy clods. If the mango plantation has to be under taken on bushy / minor fruit land, clearing of bushes and forest stuff is essential before deep ploughing. Under hillocks and step sloppy land (30-40%) slope, contour trenching instead of deep ploughing should be done. This operation provides congenial tilth to young roots for their healthy development. After harrowing, proper leveling of land is done and a gentle slope is provided in one direction to facilitate irrigation as well as drainage of excess water during rains. The soils which have drainage problems, should be provided with adequate trenching from the very beginning to avoid serious damage to young plants due to water stagnation.

Proper layout of an orchard is necessary. After field preparation, one should move for layout. Similar to selection of site, any mistake committed during layout cannot be rectified later on. Therefore, the aim and objective of plantation should be very much clear before the actual layout in the field. The system of layout and distance of planting is decided according to needs. Among the prevalent systems of layout (Square, Rectangular, Quincunx, Hexagonal and Contour), square system is very much common.

Laying out an orchard on level land is a simple matter of establishing a straight baseline, usually next to a fence or roadway. Then, lines at right angles to the baseline are established at both ends of the plot and one or two places in the middle. An easy way to establish these angles is to use three ropes whose lengths are in a 3:4:5 proportion (based on the Pythagorean theorem).Put the 40-foot rope along the baseline, then place the 30-foot rope at approximately a right angle, and, finally, close the triangle with the 50-foot rope (Fig.-3).Adjust the 30-foot segment in either direction so that it just touches the end of the 50-foot piece. This ensures that the 30-foot section is at a right angle to the baseline. Next, place stakes along the baseline and the right angle line for sighting to extend these lines. From this point on, any desired row and tree spacing can be established using a tape measure or knotted rope to measure off the proper intervals.

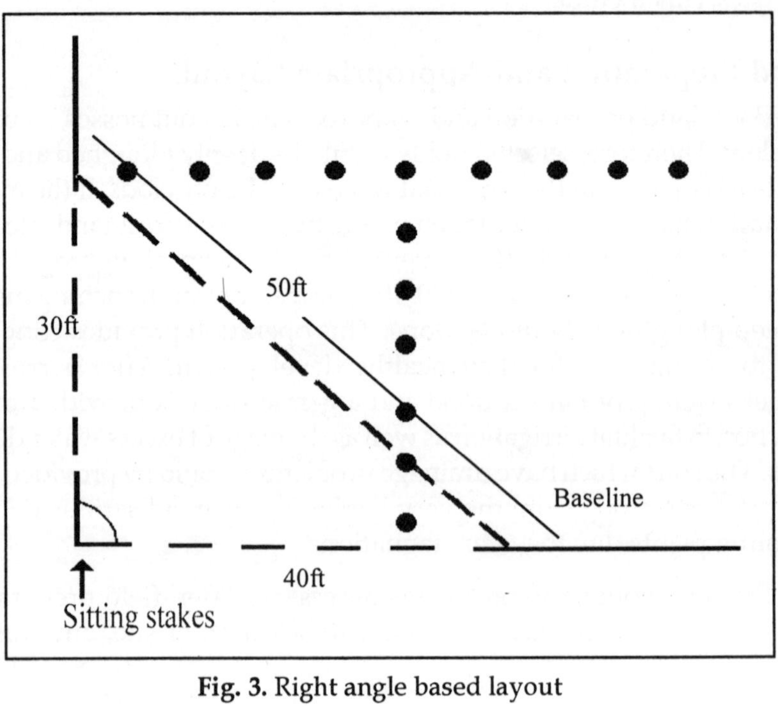

Fig. 3. Right angle based layout

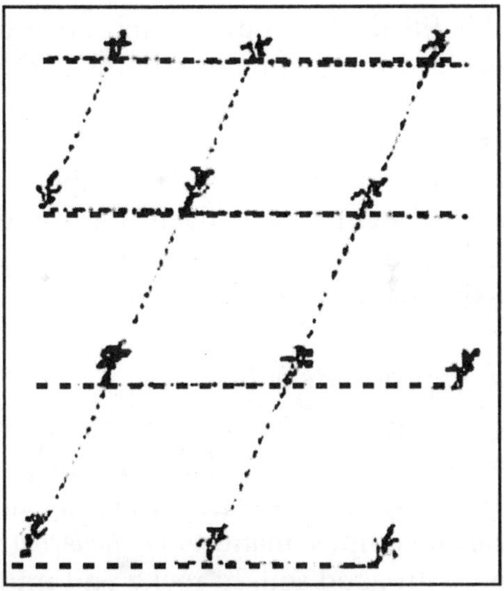

Fig. 4. Square planting

Standard Size Pit and its Filling

Dig the pits based on the soil type and other factors

The main purpose of digging and filling the pit is to provide congenial conditions for plant growth and development, specially during the establishment phase. The pits should invariably opened during summer to expose the inner surface to heat and sun light. Pits should be filled before the rainy season for maximum settling of loose soil. To prepare the pit certain steps has to be followed.

In loose soils 60 x 60 x 60 cm planting pits are sufficient while in hard soils 90 x 90 x 90 cm pits should be prepared. In morrum soils conditions, however, where good soil is lacking, pit size may be increased up to 200 x 150 cm and pit soil is replaced by good soil. In locations where the soil is loamy and deep, pits of 0.5 x 0.5 x 0.5 m may be dug at desired distances. Whereas at hill slopes, the pits should be of 1 x 1 x 1 m size along the contour. At sloppy places, the pits are prepared in form of trench by cutting the upper side of slope and leveling the lower side of the slope. In this way a leveled platform is developed near the pit. Pit should be made in proper alignment so that saplings can be planted in the centre of the pit.

When the soil is taken out of the pit, the upper half soil is kept on one side and from lower half on other, under normal soil conditions; however, in murram or problematic soil the whole soil of pit is kept aside and top soil is filled. Dug out soil and open pit are allowed to weather for 2-4 weeks during summer months to destroy any type of infection under sunlight and heat. In stony soils, it is better to separate all the stones from the excavated materials and left over soils should be mixed with manure and scrapped soil available in the field. Using a JCB machine for pit digging specially in hard and morrum soil is advisable because the pressure exerted by bucket of JCB cracks the hard pan which allows the mango roots to penetrate.

Fill the pits with appropriate substrates

Initial establishment and vigour of the mango plant is governed by the available substrate in the pit. Therefore, the pits should be filled with good quality substrate. Substrates mainly include farm yard manures, cakes, vermi-compost manure, *Trichoderma* rich manures and biofertilizers along with good soil. In rocky and murram soils, which are almost zero in their inherent nutrient capacity, the initial substrate supplied in the pit governs the performance of mango plant. As a common dose, 30-40kg, *Trichoderma* enriched FYM, 2 kg neem / karanj cake, 25g SSP or bone meal is sufficient for one pit in normal soil.In problematic soil however, pit soil should also be replaced with good soil.In highly acidic situation, application of 1-2 kg slaked lime, 250g each of Zinc and Boran (Agriculture grade) is helpful to restore initial vigour of plant.Pits should be filled one month after opening before rain.If the soil is having infestation of white ants, fenvulrate dust is mixed with upper as well as lower soil of the pit. The upper soil of pit along with substrate is filled first followed by lower soil mixture. High quality soil mixture improves initial establishment and growth of mango saplings. The main purpose of filling rich mixture in the bottom of the pit is that the plant will first proliferate their roots in 2/3rd lower portion of the pit.The upper level of pit is kept 15cm above from the field level. After filling, if rain is not there, the pits should be irrigated to settle down the soil.

Select Healthy and Genuine Saplings

The establishment of a mango orchard is a long-term venture, so it must be undertaken with great care. True to type, healthy and vigorous planting materials have a direct effect on the success of the mango orchard. Healthy planting material of genuine quality makes the base of orchard efficiency and profitability. The quality of trees grown in the nursery will determine their performance when planted in the field.

Therefore, the orchard must take off with a good start by buying high quality nursery seedlings.

Planting materials should be purchased form a registered nursery to guarantee the authenticity as to the variety and hygiene of plants. Plants can also be produced by the growers adopting the standard procedures well in advance so that planting materials attain appropriate stage by the time of planting at the orchard. Under such conditions, reliability of scion must be ensured. Selection of genuine mother plant and scion has great impact on orchard efficiency therefore, every care must be taken to select the mother plant and procure the bud wood for nursery or orchard establishment. The grafts prepared from poor, stale or diseased mother plant give poor health resulting into lower production and poor quality. On the other hand, mango grafts of desired cultivar procured from genuine sources show better orchard efficiency and production.

Buyers Guide for Mango Saplings

The buyer must visit the nursery and examine the trees closely to see if they are of a good standard. Select good quality plants by examining the following:-

Inspect mother orchard -the health and genuineness: Before purchase of planting material one should inspect the mother orchard for its health and genuineness. Saplings should not be purchased from nurseries which do not maintain the high yielding and healthy mother plants for propagation. It will be still better to select own mother plants and prepare grafts from them.

Don't purchase if over aged: Grafts which have been in the nursery only for 1-2 months should not be purchased, as there are fair chances of their mortality. A good graft should be straight and established for a year in the nursery. Too old sapling should be avoided.

Sapling height : Height should not be more than 1 m.

Leaves give clue of sapling quality: A first-grade plant should have a few older leaves on the rootstock below the graft union which can be removed if the top foliage is sufficient for photosynthesis.The leaves of a nursery seedling should have a glossy, green colour giving a general appearance of health. New leaves may have a coppery -red colour and these should not be mistaken for a deficiency. If the leaves are deformed or discoloured then there is something wrong with the tree. Trees may also be stunted due to incorrect nursery practices that result in damage by pests, diseases or mineral deficiencies.

Malformed material should be discarded : If the seedlings are flowering in the nursery, make sure that flower malformation symptoms are not present. Even after the seedlings are transplanted in the orchard, monitor them constantly to ensure prompt action if flower malformations are noted. Young trees that show flower malformation are infected and should be removed from the nursery or orchard and burnt.

Internod length: Long internodes are a good indication of a vigorous, healthy tree. Avoid seedlings with compressed internodes and a rosette of leaves at the shoot apex. Seedlings having compressed internodes are normally deficient in zinc.

Uniform graft-union: Plants should have a uniform appearance and should all be grafted at approximately the same height. An abnormal union indicates incompatibility, or a poor rootstock or a diseased scion. If seedlings show signs of bark cracking on the rootstock then it means than mono-embryonic seedlings were used and the resulting tree will be weak. Excessive swelling at the graft-union is also a sign of incompatibility and future problems. The seedling should be discarded.

Potting mixture in the sapling bag: The soil mixture plays an important role in the health of plant. Mixture with high clay content limits drainage and results in poor root respiration and growth. A mixture without essential nutrients especially phosphorus, results in poor root development. A good soil mixture should have sand FYM and soil to ensure good aeration and drainage. The best mixture should have well rotten compost, sand and loam to ensure good plant growth.

Seedling bag size and root system : Small containers result in limited root growth. The tree may appear healthy, but restricted root growth in the nursery will greatly affect growth in the orchard. Polythene pots of at least 175 x 150 x 450mm are recommended for use in mango nurseries. The suitability of the container and root system can be tested by purchasing a few trees and removing them from their containers. The taproot should be strong and straight and there should be abundance of adventitious roots.

Avoid goose neck saplings: Coiled roots and curved collor portion of rootstock/ sapling leads to restricted root growth and uptake of nutrients. Due to damage tape root, these do not go deep and deep. Some times it also leads to death of plant even after 5-6 years when water and nutrient requirement of plant beome more.Such plants should not be purchased. It can be very well detected at nursery stage.

Fig. 5. Goose necked sapling

Hardening the Sapling Before Planting

Use only very vigorous plants for field planting. Minimize the stress during field planting by hardening plants exposing to direct sunlight and with less water application. This hardening improves of the success rate of field establishment.

Ensure Minimum Root Disturbance of Grafts During Transport and Planting

Nowadays, mostly the mango plants are propagated in poly bags and plant becomes ready for planting within one year. These saplings do not have much disturbance in root system. However, the plants raised in nursery beds by raising rootstock and then grafting after one year or planting inarched plants after detachment from the mother trees face great problem of root damage. Plants in poly bag are easy to carry and transplanted with minimum mortality, however, grafts raised in nursery beds should be dug out carefully with minimum injury to the roots specially to the taproot and transported safely up to field. During digging it should be kept in mind that earth ball is clean, intact, medium in size having about 1-1.25 kg weight and packed properly with grass or straw. Heavier and bigger earth ball is likely to break and require much space for transit. The wrapping of earth ball plays important role in transportation. If loosely wrapped, chances of breaking of ball are more pronounced as compared to tightly wrapped. The earth ball should be tied at collar portion as well as in middle of the ball. Such type of tying avoids breaking of ball during transit. For transportation, grafts should be packed in medium size baskets accommodating 10-12 grafts in each basket. Before packing, the dried grasses are put at the bottom of basket which works as cushion and saves ball from jerk. Plants can also be packed in plastic crates. After making cushion the grafts are put in basket and they are tied together with string. Finally the gunny bag piece is placed over the earth ball portion of the basket. Such packed mango grafts are transported to any distance by road or train without any damage to the plants. This is for plants with earth ball whereas the plants raised in bags can be easily transported in the plastic crates.

If the distance is long and more time is required in transit, the grafts are sprinkled with water on the way and are handled gently during loading and unloading. After receiving the grafts at the particular destination, it should be kept under the shade and sprayed with water before opening of packing. Grafts with sound and intact earth ball should be planted in the field and damaged ones should be put in the nursery for hardening. Generally 5-10 per cent more plants is needed than the actual requirement. These additional plants are kept reserved. Plants should preferably be planted out before the root system out grows the containers. This will ensure the presence of strong taproots that can mine lower levels for water and nutrients.

Scientific Planting Technique

After a shower, mango is planted in the centre of the pit by removing the ball covers. The side soils of the plant ball is pressed hard to evacuate the air from the pit which otherwise restricts the root growth and then the plant performance.

Start planting at appropriate time: The best time for planting all over India is during the monsoon when there is sufficient moisture in the atmosphere. However, under North Indian conditions mango planting should be done preferably during rainy season (July to August). Planting can also be done during spring season (February to March) where assured irrigation is available. The spring season is short and followed by a hot dry period of low humidity and hence, high percentage of mortality of young plants is observed in the field. In South India, however, the main season of planting is from July to December. The best time of planting mango in the heavy rainfall zone is the end of the rainy season. In tracts where the rainfall is less, the planting can be done in the early part of the monsoon for better establishment. If it is planted during bright sun shine hours of day, it turns out to be unusually hot and dry and the plants may wither due to excessive loss of water. If the sky is overcast, planting can be done during day time also. It is better to plant the saplings in evening hours when the high humidity prevails in the atmosphere.

Table 6 : Sapling requirement at different planting density

S.No	Planting	No of plant /ha distance (mxm)	No of sapling to be procured *
1	2.50 X 2.50	1600	1760
2	3.00X3.00	1111	1222
3	4.00X4.00	625	688
4	5.00X 5.00	400	440
5	7.00X 7.00	204	225
6	10.00X10.00	100	110
7	12.00X12.00	69	76

* 10 per cent more sapling are needed to meet out mortality

Plant the saplings at proper distance: The main purpose of planning and layout of an orchard is to provide adequate space to the plant for normal development, proper intercultural operation, easy passage of air and sunlight and orchard sanitation. Distance of planting in mango is need based and dependent on various factors like nature and fertility level of soil, vigour of the cultivar, general growth conditions of the area and purpose of orchard establishment. In poor soils mango plants makes slow growth, so require less space while in heavy soils plant grow vigourusly, so require more spacing. Chausa, Langra, Dashehari, Bombay Green, Fazli are the vigorous cultivars where as Amrapali, Mallika, Totapuri, Red Small, Banglora, Alphonso, Kesar are comparatively medium or small cultivars. As such, the vigorous cultivars should be planted at 10 x 10m spacing or even more depending upon the soil types whereas the medium vigorus cultivars should be planted at 5 x 5m or 7.5 x 7.5m spacing. For high density planting 5 x 5m spacing is the best with almost all the cultivars except that of vigrours. As matter of fact, the vigorous varieties are not suitable for high density plantations. The varieties like Amrapali can also be planted little more closer at 4 x 4m or 3 x 3m 0r 2.5m x 2.5m with proper canopy management under high density system. Climatic factors of a place should always be taken into consideration while deciding the distance and planting of mango.

Adopt right method of planting of grafts : Dig a slightly bigger hole than the size of earth ball at the already marked point to accommodate the root-ball. Cut around the bottom edge of the polythene and remove the intermingled roots by pruning .Put the plant with its intact ball of earth intact with the wrapping material in the hole in such a manner that the plant remains straight and collar portion of plant is in the level or slightly higher than the ground level. Then remove the polyethylene bag with two longitudinal cuts from bottom up. After removing the polyethene cover and placing the plant in the hole, fill the planting hole with soil and slightly tighten the soil by pressing the moist soil of the pit all around the root ball so that the plant is set firmly in the field. These steps help to reduce root damage due to breaking and splitting of potting media block. The right place of planting saplings in the pit can be done with the help of a planting board.

While planting, one should be careful that the earth ball does not break during pressing the soil and the graft union remains well above the ground level. A small basin is then made and immediately after planting irrigation should be done. The frequency of irrigation in the first month depends on the weather. Planting should not be done so deep as to bury the graft-union in the soil or so high as to expose the upper roots. In any case, the scion should never touch the ground while planting or irrigation in the basin. Allow the plant to grow directly up. Use a stick closer to the plant and tighten it into the stick carefully. Regular checking of plants is essential to detect the faults like sinking of soil, tilting of plant and cracking in basin of plant after planting the grafts. Sufficient moisture should be maintained till the plants do not set or start new growth in the field.

Arrange sufficient water for irrigation: After planting, watering is an essential requirement. Water is a crucial input for establishment of mango orchard. Assured availability of water and appropriate irrigation system is the most important aspect in a commercial orchard. The type of irrigation system can be decided on the basis of water availability and capital budget. Flood, basin, check basin and furrow are the common methods, however drip method can also be used for mango orchard establishment. Under tree, drip system is the best on account of water saving but the requirement of power, technical expertise and high initial cost market the most expensive system as well. In the upland, with less irrigation facilities, mango plants can be raised by providing life saving irrigation from natural water collected in plastic lined pots / Jal kundas / open ponds. Under these situations, practices of *in situ* water harvesting and moisture conservation near plant root zone should be practiced.

Apply Mulch and Provide Shade

Use a mulch around the plant using easily available mulching material such as dry grass. Mulching helps to reduce soil temperature in the root zone. Weed control also become easy. It also reduce drying of soil and wind erosion of soil. Provide shade appropriately to protect plants from heavy sunlight.

Make Provision to Drain Excess Water

Drainage of excess water is a serious problem of mango orchards raised in heavy soils or low lying flood prone areas. In these areas, provision of proper drainage is essential before orchard establishment. Poor drainage is reflected by the poor growth, sickly appearance or death of the young plants. Building up of a persistent water table may damage root growth and the functioning of roots is affected. High water tables and high soil moisture at critical stage of plant establishment and production cycle have been found harmful. Water logging affects growth of young plants and flowering in grown up trees and subsequent productivity of orchard. The main objectives of a proper drainage system is to remove the excess water from the active root zone during rainy months and as far as possible to keep the water table in the orchard below the active root zone, that is 1.80-2.40 m below the ground level. Soils with too much of slope are prone to erosion and too much of drainage in such soils also get moisture depletion quickly and require frequent irrigations.

Plant Live Fence and Shelter Belt as Wind Break

Encroachment of tresspassers, theft and animal grazing are the matter of common concern in mango orchard. The fences act as wind break also if grown by using live plants particularly in the areas where cold and hot winds are the problem. Therefore, fencing is an important aspect in mango cultivation.

Windbreak trees are commonly tall having dense foliage and keep the surrounding atmosphere humid. These trees help in minimizing the wind velocity, low or high temperatures which affect adversely the young plants and immature fall of the fruits.

Importance of Live Fencing and Wind Break

The systematic planting of live fences and windbreaks on a farm brings the following benefits:

- Apart from high loss of water from leaf and soil surface during windy days, the wind causes fruit drop and scratches on the mango fruit surface. Harmful fungi and bacteria can enter the fruit through these wounds and make the fruit un-marketable. Damage by wind can be minimized by avoiding very windy areas, pruning the non-bearing flower panicles as soon as it is evident that these will not bear any fruit, because when they become dry and hard, they cause scratch marks on the fruit. Establishing wind breaks such as artificial structures or fast-growing trees on the upwind side of prevailing winds helps in reducing the wind damages.Wind speed is greatly reduced, preventing physical damage to flowers, fruit and plants by dust, hail, snow and abrasion. Both the quantity and the quality of the crop are improved by the planting of windbreaks.
- The main purpose of fencing is to protect the plants from stray animals and harmful effects of cold and hot waves. Besides protection, the main function of live fences, to provides fuel wood, fodder and food, and enrich the soil,
- The pernicious damage by wind erosion to the soil resource, which results in a lowering of soil fertility, is reduced,
- Lowering wind speed reduces the chill factor, resulting in higher stock survival,
- Lower wind speeds reduce evapotranspiration, resulting in more efficient use of available soil moisture,
- Low wind speeds facilitate pollination by bees and other insects,
- Shelter is provided for desirable birds and animals,
- Protection from wind prevents lodging of mango crop,

- Windbreaks can supply products such as timber, poles and firewood and provide nectar and pollen for honey production, and
- Windbreaks add to the attractiveness of a farm.

Disadvantages of Live Fences and Windbreaks

- Windbreaks planted on the contour can trap cold air resulting in frost pockets), or hot air (resulting in heat scorch), both of which can cause crop damage.
- Windbreaks running east-west will shade the southern side all day, reducing crop growth and slowing down the drying out of roads and poorly drained lands after rain. Land which would otherwise be under crops is taken up by trees.
- Tree roots can affect the productivity of the adjoining land for a distance of approximately their own height. This effect can, however, be greatly reduced by annually ripping a single deep line parallel to, and 3 to 4 metres away from, the tree line to keep tree roots away from the crop.
- Undesirable birds, such as finches, find shelter in windbreaks.
- Too dense barrier can result in winds becoming turbulent in the lee of windbreaks, causing physical damage to crops.
- If the wrong tree species is chosen it may invade adjoining areas.

Fencing is done in two ways; fencing by using cement pole and wire or by erecting boundary wall and live fencing by growing of specific plants species. The first method of fencing is costly and can be afforded by rich farmers only. Live fencing is an age old and traditional practice. Traditional farmers have vast knowledge about the practice live fencing and the species used. Plating live fence with tall growing trees as first row and small, bushy thorny plants in staggered manner as second row along the boundary of mango orchard is very much important. In sloppy lands wind breaks should be planted along with contour in ascending and descending order to protect the plants even at higher points. At such places, the live fencing can not be combined with wind break and must be planted along the boundary. Having a live fence around the farm has multiple benefits.It does however, take two to three years to develop. The species suitable for live fence should be thorny, inedible and non-grazable for cattle and goats, hardy and

Fig. 6. Live fencing

relatively maintenance free (other than pruning / lopping), adaptable to the local conditions, fast-growing with ability to produce something that can generate some revenue. Live fences can be divided into two basic categories on the basis of their growth habit and use.

How Does a Windbreak Work?

When wind approaches a windbreak, part of the moving air filters through the leaves and branches, reducing its speed and form an air cushion on either side of the windbreak. The remaining air is deflected over the windbreak and the leeward air cushion helps to keep it away

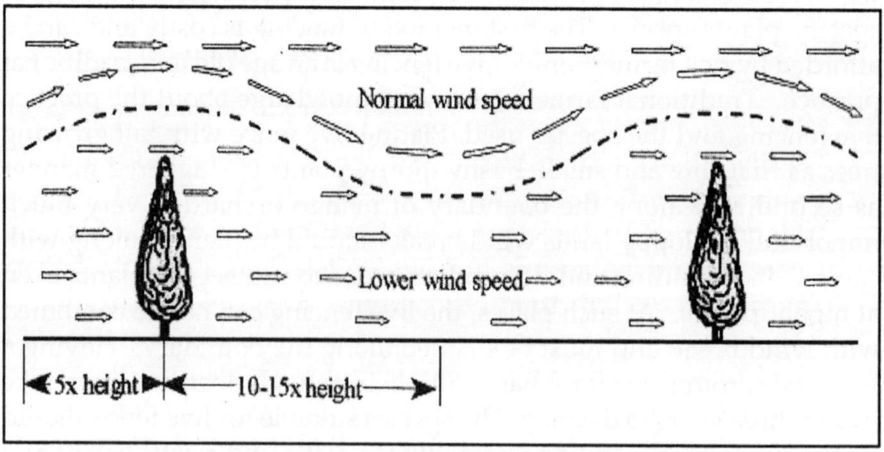

Fig. 7. Wind movement as affected by wind break

from ground level for a considerable distance. The area of reduced wind speed may be up to 5 times the height of the trees on the windward side and up to 20 times the height on the leeward side, depending on wind speed, topography and other factors. A windbreak 10 m high can provide shelter over a total distance of up to 250 m, but a closer spacing is desirable to prevent the wind speed building up to more than half its original speed. The height of the crop must also be taken into consideration as this reduces the effective height of the windbreak. For example, with a 10 m high windbreak and mango trees 4 metres high, the effective height of the windbreak is only 6 metres.

When the wind speed is reduced by one quarter, its force is reduced by approximately one half, greatly reducing mechanical damage by abrasion.

Live Fence Posts

Fast growing hardly plants species can be grown as fence post at a spacing of 4-5m.These plants commonly grown by seeds or maintained from natural vegetations growing along the boundary to have strong root system. Farmers can also plant stakes of easy to root species such as *Gliricidia sepium, Moringa,* Flame tree etc. The live fence posts are more durable than traditional wooden poles as they are less prone to attack by termites and decay fungi. When grazing animals are part of the farming system, the only way to establish live fence posts and a living fence, is to start with a conventional wire fence supported by wooden pole and to gradually establish live fence plants to substitute for the decaying poles. *Gliricidia* live posts fence can be established by planting a few large (1.5-2.0 m) stakes in the existing conventional wire fence. These stakes normally take root within a month or so. After the first pruning, subsequent prunings can be carried out every 4 to 8 months. Shoot pruning at intervals of 6 to 8 months results in woody sprouts that are suitable for use as stakes. These pruned twigs can be used to multiply live stakes for fence posts within a year or two after establishing the first live fence posts. Tar, Babool, Simaruba, Jamun, Jangal gilebee etc. can also be planted as fence posts.

Live Barriers or Hedges

Hedges are thicker, more densely spaced fences that generally include a number of thick growing thorny species which do not require support of barbed wire. To minimize the initial cost of mango orchard establishment, it is advisable to establish dense and thorny hedges to protect the orchard. Another alternative often used by farmers is the combination of easy to establish live fence plants that are poisonous and unpalatable species. If well established, these natural barriers can deter both animal and human trespassers from entering into the orchard. A live fence should ideally be planted just before the monsoons and watered regularly after the rainy season is over to ensure optimum growth. Agave, Sisal, Boradi, Karonda and Mehandi etc. are the promising plant species which can be planted as live hedge along the mango orchard boundary.

The trees, which are generally used as windbreak, are seedling mango, jamun, mulberry, moringa, jackfruit, carambola, shisham, teak, gamhar and bamboo. These can also be used as live fencing posts. These trees are usually planted close to each other to provide an effective shield. Tall growing trees like shisham and jamun are planted 6m apart while low headed trees like mulberry and carambola are planted at 7m distance. Normally, plants grown as wind break may extract nutrients from mango plants, at least from the boarder one or two rows therefore, a trench of one meter depth should be dug between mango row and wind break.

❑❑❑

6

Use in-situ Orchard Establishment Technique in Hilly, Sloppy and Adverse Land Conditions

Generally, upland farming systems are based on production for subsistence, and include extensive shifting cultivation, permanent cultivation, home gardens, grazing and any combination of these. Over time, there has been a tendency towards intensification of land use and permanent upland cultivation. Mostly such farms are not accessible by road and lying kilometers away from roads. There are vast areas in many states where other agricultural crops are not that much economical as fruit crops. Traditionally, many fruit crops are being grown under sloppy, hilly and adverse situation. For successful mango cropping in such areas, the technology may differ from plain land.

Land Development

Heavy rainfall and poor vegetation cause severe erosion in sloppy land. Therefore, orchards are protected by building slope-specific soil conservation treatments. Trench-cum-bunds are constructed on land with slopes less than 15 %. Where slopes are more than 15%, trees are first protected by platforms and subsequently overall plot treatments such as trenching, gully plugging, and terracing are applied.

Terrace is built by taking earth from the top of the slope and moving it down to an area farther down. This allows turning slopy fields into flat fields. Once the terrace is ready, a low earthen wall about 6-10 inches high is built around it to increase the amount of water that enters the ground. One or two gaps are made in the wall to prevent water logging of the field. Water is drained away from other agricultural lands at down stream to prevent damage. If slop is low terracing may not be needed. The mango saplings are planted on hill sides and sloppy land across the slope instead of up and down the slope. This helps to keep valuable topsoil in place, slowing water down and it's soaking into the soil, improves irrigation systems and conserves water, reduce labor and make harvesting easier and improves the appearance of the farm.

Orchard Layout

A place on the slope is selected to start with. Contour line is made across the slope with sticks or stones. A hole is dug and walked about 1 m along the contour line, another hole is dug. This process is repeated until the edge of the field is reached. Then a point is marked at about 10 m (30 feet) up or down the slope. Thus, another contour line is marked measuring the spacing. Same thing is done again until the required numbers of marks are obtained on contour line. Pits are made in the beginning at the steepest point of the field (fig. 8), and rows follow the contour of the land. First row is made at the highest elevation and stakes are fixed (i.e., to make all points on the line the same elevation, A to C. Next, the steepest slope along this row is selected (along the line A to B) minimum distance between rows is measured. From this point, next row is laid out on a level line as before. As one moves from the steepest slope to less-steep slopes, the rows become wider apart. Wherever the distance between two adjacent rows becomes twice the minimum distance, a short contour row is laid out between them from that point to the end of the plot (squares marked D).

Pit Digging and Filling

Pits should be of 1 x 1 x 1 m size along the contour. At sloppy places, the pits are prepared in form of trench by cutting the upper side of slope and leveling the lower side of the slope. In this way a leveled platform is developed near the pit. In hilly areas, there is shortage of good soil and many times pit contains useless materials. Under such situation, the top soil is collected from the surrounding to fill the pit. The pitting materials may the same as used under normal situation.

In-situ Planting of Rootstock

Rootstocks affect plant growth and tree performance. Besides, these are helpful in growing plants under adverse soil and climatic conditions.

Generally, mango stones are collected and dipped in insecticide solution (endosulfan 0.07%) for a few minutes. Use of fungicides such as 'Captan' at this time is helpful to protect the seeds from fungal infections. Then at least 4 seeds are planted at one place in the middle of the pit. Care should be taken that the moisture is optimum for seed germination. After placing the stones, the moist leaf mould is placed over them. The seeds germinate within 20 days after sowing. These may be irrigated time to time. Spray of chemicals may be done to protect young seedlings especially from leaf cutting weevil and diseases. If seed bed is containing materials such as coir dust, care must be taken to maintain the appropriate moisture level and management of termites.

In -situ Grafting

The seedling become ready for grafting with in one year. The saplings raised from 5 different methods (inarching, budding, air layering, veneer grafting and stooling) when grown in the nursery for 20 months differ in their growth capacity. Establishment of a grafted mango orchard of uniform trees is somewhat difficult when done by nursery grafted plants. There is considerable mortality, if the nursery nursery grafted plants are put in the field. This involves a lot of periodical replacement, resulting in a heterogeneous group of trees. Besides bad appearance of the orchard, these different trees take different time to produce an economic yield. Vigorous seedlings are obtained in *in-situ* planting and these may be veneer/cleft grafted with scion sticks of the

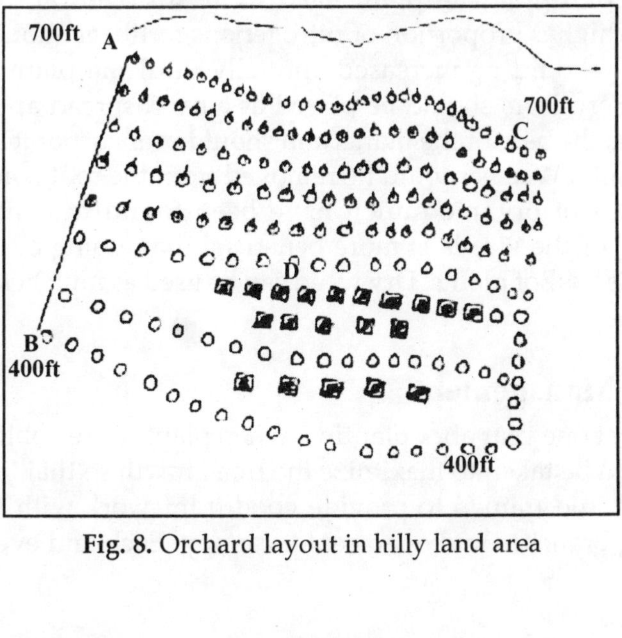

Fig. 8. Orchard layout in hilly land area

variety one desires to plant. All the care should be taken in selection of scion as in the case of normal grafting. Such plants grow very fast and attain a stature in a few years, which the inarched plants would have taken many more years to attain. To prevent drying off of the scion wood, cover it with a polyethylene bag. In case, there is mortality, adjoining another rootstock may be used for grafting. After success of graft, the rest of root stocks may be removed.

Care and Management of Plants

Proper aeration near root zone and good sanitary conditions in the basins is a must for keeping the plants healthy and disease free. Proper growth of young plants can be ensured by timely hoeing and weeding of the basin which is considered to be the essential. It is necessary to protect the sapling from cold as these are more exposed to hot and cold waves. Protection of mango saplings for at least one winter or in some cases two seasons is very essential. For the purpose, thatching of individual plant is more essential. In early stage, inflorescences should be nipped off immediately after emergence so that it may not disturb vegetative growth of the newly established mango plants. There may be less need of irrigation in such orchard, however, in summer, proper moisture is needed to support the scion otherwise it will dry leaving rootstock alive. Therefore, every effort should be made to maintain sufficient moisture near root zone of the plant. The quantity and frequency may have to be modified depending on the soil type and climatic condition The plant need more nitrogen at young stage therefore, higher proportion of nitrogenous fertilizer should be given. The doses should be increased annually until the plant attains full growth. Fertilizers should be placed as a band spread approximately 30 cm from the base of the plants and should be incorporated to the soil mechanically. Weeds should be removed from the root zone of mango plants.Uses of organic mulch have been found to control weeds. Chopping of the weeds is more beneficial than doing clean weeding around the base of plants. Dry weed can be used as mulches around the trees.

Canopy Management

Almost one year after planting, when plants are established, every care should be taken to maximize the tree growth so that the plant can be pruned and trained to provide good framework with well spaced spreading branches and open center canopy. Each and every plant on hill or sloppy land experiences more wind velocity in comparison to plants in plains for the reason that they are located at different height. Therefore, every efforts should be made to keep plant stature dwarf and manageable to avoid wind damage. The technique of canopy management has already been mentioned in this book.

Utilization of the Interspaces for Cultivation

Land surface should not be kept exposed all the time as water current may erode the soil. Therefore, clean cultivation throughout the year is not desirable in such plantation. In rainy season, no ploghing should be done or if at all some crop is to be sown it should be done at the start of rain only. If nothing is to be grown, orchard grasses should be allowed to grow thoughout the rainy season which later on can be used as mulch materials. Growing fodder grass is the best practice in such mango orchards. Some intercrops may be grown in vacant space between the trees. The selection of crops should be purely area based and as per need of the farmer preferably, the leguminous crops.

Saving the Young Plant from Enemies, Pests and Fire

Bio-fencing should be done on sides of orchard to save the young orchard from cattle and wild animal. The method of raising bio fencing has been given in this book in chapter-5 which may be referred for the purpose. Similarly, the measures to control the pests and diseases should be taken well in time to save the developing mango plants. Most of the orchards in such situation remain fallow during rainy season. If growing grasses are not chopped, they posses threat of onset of fire which may damage the mango. Therefore, the grass management, especially when they are dry, is very important. If grasses are not managed, a vacant strip between the rows may be left or a clean circular ring may be made around the basin to protect fire.

Soil Erosion and Water Management on Sloping Orchards

These areas experience two extreme situations - heavy rainfall during the monsoon and water scarcity during summer. Therefore, activities for harvesting available water and its efficient utilization have been developed. One should be aware of water movement and should be extra careful to protect orchard soils from water damage. Planning may be done for heavy rain before the rains actually fall. Long, smooth and even slopes speed up runoff water. Fast moving water has a lot of power to cut deep into the land. It can wash soil off field and damage or even destroy the trees. Attempts should be made to shorten slope lengths by making them more irregular, for slowing down rainwater. Controlling runoff and rebuilding the soil on sloping land is a difficult work. Degree of slope, ground cover, rate of water percolation into the soil and into the subsoil are some considerations that decide the degree of effort to be made for soil conservation in sloppy orchard. In the first year, 1m wide shallow trench is built along the contour. Work may be started at the highest point of the land and water may be controlled there first and then working the way down the slope putting appropriate design into effect. Water is carried specifically to a dam, pond or tank, sinking water (allowing it to seep into the soil) which is the aim of all water management. Some of the prevalent methods are given below.

Hillside Ditch

Hillside ditch is made by digging a small ditch across the slope to divert rainwater. With some basic earth shaping, one can install these hillside ditches across the slope. This will shorten the slope length and divert water to a stable area on the farm. With this additional practice, the farm is better protected from water damage.

Vegetative Barriers and Conservation Cover

First step for slowing down rainwater is to determine field's contour lines. Planting some cover crops across the slope or on the contour and growing small strips of stiff plants across the slope across helps to shorten slope lengths, slowing down runoff water so it can soak into the soil. One can plant conservation covers, vegetative barriers, and alley cropping hedge rows in strips or rows across the slope.

Mulching and Residue Management

Use of mulching and residue management is done to keep the soil under the trees covered. Plant residue (such as banana stalks and tree trimmings) can be placed in strips across the slope to help trap and retain water.

Irrigation Practices

Drip irrigation

Three types of irrigation systems are commonly used are surface, sprinkler and drip irrigation. Drip irrigation is the most efficient in terms of water use in such areas. It provides good water control by delivering water near the plant, enabling the farmer to grow crops with much less water than with other methods. In drip irrigation, water flows through a filter into special drip pipes, with emitters located at different spacing. Water is discharged through the emitters directly into the soil near the plants through a special slow-release technology. Drip irrigation is especially useful for undulated and sloppy land. For advance drip irrigation filter, drip tape or polyethylene pipe, drip emitters, connectors, water source are needed. To operate the system, electricity and motor are also .However, this is a costly affair.

Temporary Check Dams

Immediately after the monsoon, the run-off water is harvested by constructing temporary check dams across the nalha and streams. These are constructed inexpensively by using empty cement bags filled with sand and silt.

Spring Development

A number of well dispersed perennial springs exist in the hilly areas. Though ground water resources are poor, the springs provide sufficient water to irrigate the orchards and cultivation of mango crops on a small scale.

Pot Drip Irrigation

Pot drips are used in order to reduce the drudgery of irrigating orchards from head-loads of water and to reduce losses from evaporation and infiltration. The pot drip system consists of four clay pots dug in around the plant. The water in the pot diffuses into the soil through a hole in the bottom of the pot and in this way becomes available to the plant. This can result in water savings of over 50 per cent.

❑❑❑

[illegible] near the plant, enabling the farmer to grow crops with much less water than with the furrow flood and sprinkler irrigation [illegible] a [illegible] When [illegible] plants [illegible] especially useful [illegible] and [illegible] supply and [illegible] water [illegible] are needed to operate [illegible] are also [illegible] so this is a costly affair.

[illegible]

[illegible] and [illegible] the [illegible] water [illegible] sand and salt.

Spring Development

A number of [illegible] dispersed pockets [illegible] Thal [illegible] groundwater resources [illegible] springs [illegible] water to irrigate the orchards [illegible] scale.

Pot-Drip Irrigation

Pot drips are used [illegible] around the plant. The water [illegible] hole at the bottom of the pot [illegible] plant. This [illegible] 30 per cent.

7

Maximize Profit Through High Density Planting and Orchard Floor Management

Advancing knowledge in mango tree architecture, growth physiology, development of dwarfing rootstocks and chemical growth retardants, has enabled farmers for closer planting and maintaining small sized trees. This system, popularly known as the High Density Planting (HDP), which enables earlier cropping, high regular yields and improved farm management practices, leading to higher productivity and profitability. Today, majority of new orchards of mango are being planted in this system with higher fruit production and increased profitably. Use of growth retardants which restricts tree growth and encourages early flower induction, has also been found helpful for these high-density planting systems. Perhaps, presently the high density farming has been the area of most intense focus in the mango production. To meet the challenge of high productivity, one has to optimize the parameters of growth and minimize the unproductive components of plants without sacrificing the overall health of the tree and quality of the product.

High Density Planting

High density planting technique is a modern method of mango cultivation involving planting of mango trees densely, allowing small or dwarf trees with modified canopy for better light interception and distribution and ease of mechanised field operations. Control of pests and diseases, weeds and pruning of tree canopy and sometimes harvesting are carried out by machine. Fertilizers are applied through fertigation. Irrigation and fertigation are automatically controlled. Such systems have been reported to give precocious cropping, high regular yields of good quality fruits and low labour requirement to meet continually rising production costs. It has been successfully adopted in the cultivation of temperate fruits especially apples and also showed its promise in mango.In the case of relatively short duration fruit crops like banana, pineapple and papaya, the advantage being that any mistake committed in spacing of these short-term fruits can be rectified easily without much financial loss. However, in the case of mango, hasty attempt for closer spacing may bring negative impact in growers' fields if the complete package of high density has not been properly understood.

Reasons for HDP as Against Normal Planting

Increasing pressure on land owing to diversion of orchard lands to other purposes like urban development and infrastructure generation as well as rising energy and land-costs, together with mounting demand for fruits have made it imperative to achieve higher productivity mango from limited space.

Most of the commercial mango cultivars in India are location specific with long gestation period and alternate bearing habit. The normal planting distance in these cultivars has been 10-12m. The orchard takes 10-15 years to provide economic returns depending upon the cultivar, planting distance and other cultural practices. Due to poor early returns and clash between the cultural requirements of the inter crop with main crop, mango orcharding so far is done by large farmers

Fig. 9. High density mango orchard

who can afford tall trees and or failure of inter crops particularly in the early stages of orchard life (up to10 years).In mango, vegetative and reproductive phases are antagonistic to each other. When the tree is full of vigour, dominance of vegetative phase prevails over reproductive phase,which adversely affects flowering, fruit set and fruit growth, besides fruit drop due to high competition for photosynthates between vegetative and reproductive phases of the tree.

The tall trees of normal density at high productive stage require heavy machinery and equipment for spraying of pesticides and crop harvest because outer tree canopy is the major fruiting area extending outwardly with the enlargement of the tree. High density orcharding appears to be the most appropriate answer to overcome low productivity and long gestation period for early returns and export of mangoes. In high density system, yields are improved in early years of orchard life.

Major Components of HDP

Planting density :Even though, a small canopy with a high number of well-illuminated leaves is efficient in photosynthesis but it is very poor in light interception, which leads to low potential yield per hectare. Light interception sould be improved by increasing tree density. An optimum tree density is the level of tree density that is required to produce optimum light distribution and interception which leads to high canopy photosynthesis. As a result, yield per hectare is maximized. An optimum light interception is a factor of plant form, planting density, tree arrangement and leaf response to light for photosynthesis. Optimum light interception can be defined as a level of light intercepted by an orchard system above or below which, the economic yield will be reduced. Depending on the planting system, tree density under high density planting ranges from as low as 400 tree per hectare to an ultra-high density of 1600 plants per hectare.

Planting system : Planting system is a combination of tree arrangement and plant form. Tree arrangement in HDP system must have sufficient alleyways for movement of farm machinery. The way trees are arranged also determines the light distribution pattern and light interception level. As of now, full field system is common for HDP. This system is called full field system because in this method mango trees are planted without alley-ways. In term of light interception, this is the most efficient system. However, the full field system is only suitable for small scale orchard due to the difficulty in field operation through machines.

Pruning system : The tree growth in low density mango remains similar as in high density mango upto eight years. The horticultural methods most commonly known to control tree growth are training, pruning, use of dwarfing rootstock and growth regulators. The training begins when the tree is first planted and continues through out its productive life. Proper tree forms, branch angle and limb spacing in themselves aid in growth control. First training is done after one growing season. Each plant is allowed to maintain single stem (main stem) with upward growth .The top of the main stem is removed by cutting out at 60-80cm height from the soil surface keeping in view that the cut is made above an extension whorl. 4-5 primary branches emerging between 60-75cm height are retained on each main stem. Thereafter, 2 shoots arising from each primary branch at a distance of 60-75cm from main stem is allowed to form secondary and like wise the tertiary branches. After start of bearing in plants, shoots arising from secondary and tertiary branches are given 5-10cm deep pruning in July soon after fruit harvest. Spray of 1% urea combined with 0.2% Blitox-50 or any other copper fungicide should be done soon after pruning in July.

The trees of high density orchard show slightly higher growth in terms of height from 8th year onwards than those of normal density because of more pressure from side plants and want of sun light. A reverse trend occurs in the growth of crown circumference and main stem girth and the differences in the growth of trees in different densities wideness after 10th year. Once the trees have filled their allotted spaces crowding may occur and canopies of adjacent trees begin to overlap. This may lead to excessive shading and reduction in photosynthesis by layered leaves within the tree canopy resulting in poor yields. In fact, at some point of time, most fruit trees require vegetative growth control, particularly in high density orcharding. Once the tree is mature, excessive growth can be regularly removed by pruning to provide a short term or immediate benefit. In high density planting, the main objective of pruning is to develop plant form which is good in light distribution, small in size and easy to manage especially in terms of mechanization.

The tree height and canopy of plant in high density orchards (1600 plants per hactare) is reduced by back pruning and dehorning of branches after crop harvest in 10th years onwards, besides light pruning of fruited shoots (5-10 cm top) in other years as regular practice. Dehorning is a process of removal of some selected branches from their place of origin on limb for vigour control and light and air penetration. Regular pruning induces new shoot production within one month during July-August. In general, every pruned shoot produces one new shoot but occasionally more than one shoot is also produced. After harvest of crop, about 50% branches of the tree canopy should be dehroned in high density orchard when the plant starts intermingling of canopies and the remaining 50% branches should be dethroned in next July after crop harvest to have a proper balance. Dethroning is done to avoid overlapping of trees in the high-density orchard. Paclobutrazol may be given to bring about dwarfing in high density planting.

Good light distribution within tree canopy increases the number of well illuminated leaves. This condition induces a high rate of canopy photosynthesis that leads to high yield/tree. Small and compact canopy plant form is desired for mango high density which can be achieved by planting dwarfing varieties like Amrapali, dwarfing root stocks and constant pruning and dehorning.

Automatic systems : Another component in high density mango planting is the system automation which contributes to high production efficiency. One of the important farm operations that can be automated is irrigation and fertigation. In fact, irrigation and fertigation have been identified as one of the key factors for the success of high density orchards in case of mango and many other fruit crops. Plant should not be kept under stress after pruning therefore, assured irrigation coupled with fertigation is essential after pruning and during fruit development in high density mango orchards.

Nutrition and irrigation :High density mango orchard should be given 100kg N, 75kg P and 75 kg K per ha every year. Soil application of full doses of P and K should be done in July; whereas N may be given in two equal doses in December and after crop harvest in July. Besides, spray of 1% urea (5 kg/ha) at the time of maturation of new flush improves photosynthetic efficiency of leaves. In case of fertigation, the standard schedule should be followed during vegetative reproductive growth phase and fruit development stage. Leaf litter should be allowed to decompose in the orchard soil.

High density orchard should be irrigated soon after paclobutrazol application during September-October, if there is no residual moisture in soil. Thereafter, the orchards are irrigated from panicle emergence to crop harvest at 15-20 days interval from March to June. However, no irrigation is applied from first shower of the rains in June to February end. Tillage operations should be done in October-November to keep the orchard free from weeds and also to mix fertilizers and leaf litter into the orchard soil.

Control of Pests and Diseases

The pest and disease problem remains similar to normal planting except some pest such as hoppers, mealy scales and powdery mildew. To control the pests and disease the recommended practices and chemicals should be followed as per recommendation. Weed cause some problems in initial stage of orchard. These should be removed and mulching may be done. In later stage the weed problems goes down in HDP.

Fruit Yield

Mango plants under high density planting system trees start fruiting in third year of their planting. If the orchard has been developed through *in situ* method, the age of rootstock will decide the bearing. In high density planting, cultivar Amrapalli safely produces a yield of 80-85t/ha within 10 years of planting.

Economic Impact

The mango trees under HDP system are precocious and start bearing 2-3 years after planting. Production per unit area is increased and early return to investment is possible. There is also better light interception and distribution, thus improved fruit quality. The increase in yield, early and maximized return, reduced cost of production and improved fruit quality result in greater orchard profitability and provide better competitive advantage of mango growers. It is appropriate to adopt complete package of high density mango production. Only planting and or pruning will not be of much use for HDP.

Orchard Floor Management

Management of ground space of orchard for additional benefits without hampering the performance of main crop is known as as orchard floor management. Mango trees take longer time to yield any profit to the grower, it is desirable to grow some short-term crops in initial stage of orchard establishment when there is no income from the primary crop and should continue till the main crop plants come in full bearing. Such crops when taken in the orchard are called intercrops. Thus, mango cultivation provides an opportunity for utilizing the land space to its maximum during initial years (up to 8-10 years) of establishment. Due to wide spacing and developing root patterns, the large unutilized interspace can be exploited for growing inter and mixed crops successfully. The soil fertility can also be maintained / enhanced by careful selection of intercrops and adequate management of the orchard. However, selection of intercrops depends on agroclimatic region, marketing facilities, levels of inputs and other local considerations.

There is another kind of cropping in the orchard when the orchard soil is covered with some crop which in due course is turned back to the soil to supplement organic matter of the orchard soil. This is called cover crop. Cover crops save the orchard soil from being eroded during the rainy season, particularly in areas where drainage is not proper. When the cover crops are turned into the soil, the organic matter of the soil is increased, and this results in many advantages. For example, if organic manure such as compost or farmyard manure is not available, the nutrient requirements of fruit trees can be met by applying only artificial fertilizers without any adverse effect on the soil. Water-holding capacity of the soil is increased and the biological complex of the soil is also improved. These two types of crops (intercrops and cover crops) are important operations in the management of a good orchard, although many growers in India do not pay much attention to it.

Consideration for Intercropping and Cover Cropping in Mango

Crop selection

- Intercrops should occupy a secondary place in the orchard, primary consideration is given to the perennial fruit trees. The crops that may grow tall and have a tendency towards excessive growth should be discouraged.
- Water requirements of the intercrops should as far as possible coincides with the requirement of fruit trees. Such intercrops should be selected that do not exhaust the nutrients and moisture from the soil which are essential for the growth of fruit trees.
- Perennial or exhaustive crops should be discouraged as an intercrop in the orchard. This may have devitalizing effect on the

growing trees. For example, sugarcane, pigeonpea, maize, *jowar* should invariably be excluded from an intercropping programme in the orchard.

- Vegetables that have their roots within 25cm depth of the soil are considered good for intercropping. These are tomato, onion, cauliflower, beans, radish, *palak* etc. These vegetables can be taken profitably in a growing orchard.
- It is preferable to take leguminous crops for cover cropping so that nitrogen fixation in the orchard soil is facilitated. Another advantage of cover crops is that they will prevent the growth of weeds. During rainy season *(kharif)* greengram, blackgram, cowpea and *guar* are preferable. During winter season *(rabi)* it is desirable to take sweet clover, pea, lentil, mothbean and fenugreek.
- Some crops like sunhemp in light soils and Dhaincha in heavy soils may also be grown to protect the orchard soil from erosion and also for enriching the soil fertility. These crops are sown in beginning before rainy season and ploughed after monsoon season.
- Crops which are cultivated during *rabi* and which require regular and good amount of irrigation should not be preferred. The reason is that the fruit-bud differentiation of mango under subtropical conditions of northern India takes place in November- December, and irrigation during this period or before will interfere with the flower bud differentiation and affect ultimately the production.
- Another groups of crops which need to be avoided as intercrops are those which are grown for seed/ grain purpose. This includes

Fig. 10. Pine apple intercroping in mango

wheat, oats, gram, oilseeds, pea and radish (for seed). The reason is that the period of seed formation and subsequent development of seed in the intercrop generally coincides with the period of active fruit development in mango, and during this period the intercrops extract maximum amount of moisture and nutrition from the soil and deplete it, irrespective of the care taken to replenish it for fruit trees.

- Colocasia, ginger, turmeric and grain crops are basically heavy feeders and extract maximum amount of nutrients from the soil for several months during which these crops stand in the field. Severe defoliation of trees is observed in the plots where ginger is grown as an intercrop,hence, such crops should therefore, be avoided. The partial shade loving crops like pineapple and amorphophallus etc. can be grown in fully grown orchards. In addition to field crops, some short duration, less exhaustive and dwarf type inter-fillers like papaya and guava can also be grown till these do not interfere with the main mango crop. These inter-fillers can be selected depending on region and other considerations.
- Leguminous crops of like greengram, blackgram, gram; etc., cereals like wheat; oilseeds like mustard, sesame and groundnut and vegetables crops such as cabbage, cauliflower, tomato,potato, brinjal, cucumber, pumpkin, bittergourd, tinda, lady's finger, etc. and spices like chillies can be successfully grown as intercrops.

To make intercropping a profitable venture for better utilization of land in between the tree rows, it has been found practically feasible to grow limited number of crops in the following rotation.

Intensive Rotation of Inter Crop

- Cabbage (Oct-Feb.) – Cowpea (March-June)-Methi (July- Sept).
- Okra (Sept.-Dec.)-French bean (Jan-March)-Tomato (April- July).
- Potato (Nov-Feb)-amaranth (March-May)-Cowpea (June- Aug.)-Radish (Sept.-Oct.).
- Cauliflower (Aug.-Nov.)-Potato (Dec.-March)-Dolichos (April-July).
- Cabbage (Sept.-Dec.)-Brinjal (Jan.-May)-Cucumber (June- Aug.).
- Palak (Nov.-Jan.)-Cluster bean (Feb.-May)-Brinjal (June- Oct.).

- Bitter gourd (Dec.-March)-Cucumber (April-June)-Sponge gourd (July-Oct.)
- Pumpkin (Nov.-Feb.)-Ridge gourd (March-June)-Bitter gourd (July-Oct.).
- Cucumber (Dec.-March)-Bitter gourd (April-July)-Dolichos bean (Aug.-Dec.)

Medium Duration Rotation of Inter Crops

Cow pea- potato, Green gram- wheat, Black gram- wheat, Green gram- gram, Black gram-gram, Cowpea-wheat, Cow pea- gram

Fig. 11. Cover cropping in mango

Annual - Perennial Rotation Crops

Pine apple, papaya, Amorphophallus.

All the crops mentioned, do not do well under dense shade, which obtains in an old orchard having large spreading trees. Intercropping under such a condition is not advisable. However, such orchards should be protected from weed infestation and soil erosion commonly seen during rainy season. For such situation, in north India cover crops should be grown in between the tree rows. Cowpea has been found very suitable. It is quick-growing, fast-spreading, broad leaved and covers the ground effectively. This crop can be turned in under the soil during September-October as it does not mature before October.

Method of Growing

The method of growing intercrop in an orchard is different in the young and the bearing orchards. In the former, the entire land is utilized, leaving only the basin portion of the tree. The tree is located in the middle of the bed in which the crop is grown. As the trees grow in age, the size of the basin increases and the area for intercrops gets reduced gradually. When the trees attain good bearing age, the intercrop is grown in between the tree rows in 1 direction, i.e. on 2 opposite sides of the trees, leaving the other 2 sides open. The tree rows are enclosed by a strong bund on 2 sides running from one end of the orchard to the other. This facilitates irrigation of trees within the bunded space independently of the intercrop. It also provides space for movement during plant protection and other operations in the orchard.

Pest Management

If the cover crops or intercrops are taken regularly, the weeds will be automatically suppressed. A constant watch over the diseases and insect pests of mango is necessary for ensuring quality crop from the trees. When the humidity builds up appreciably during the flowering time, incidence of mango hopper and powdery mildew is very much increased, and sometimes this may result in no fruit set.Likewise incidence of coccids and scale insect becomes more when intensive intercropping is done and pesticides are used regularly. Thus, intercropping/cover cropping in orchard may promote infestation of pests and diseases in greater intensity than under clean cultivation. This is due to increase in the humidity and vegetation around the trees due to intercrops. Hence, regular plant protection measures against pests and diseases of the mango crop are essential.However,no vegetable insect pests have been found to attack mango crop and vise versa in intercropping in of mango orchard.Skirting and centre opening like pruning process must be followed from time to time in mango plant for successful intercropping.

Irrigation and Nutrition

While taking intercrop, a common mistake is to leave the fruit trees to obtain nutrients from the same fertilizer dose applied to the intercrops. This is not a good practice. The intercrop and the fruit trees must receive separately their independent requirements of fertilizer and irrigation. The nutrition and moisture to intercrops must be provided in additional amount. Proper irrigation of trees, particularly during summer, is very essential to get quality crop from the bearing trees and proper growth in young plants. While bearing trees should to be irrigated at an interval of a fortnight during summer, younger plants need weekly irrigation.

Trees must be regularly observed for any setback due to any cause and steps must be taken immediately for its remedy. A proper round the year cultural schedule must be drawn up before hand and operations should be done as per schedule. This will ensure production of quality crop from healthy mango trees and good yield from intercrop.

❑❑❑

8

Take Special Care and Manage the Developing Plants

Mango saplings planted in June-July take full advantage of monsoon season for their establishment and initial flushing. Up to October, plant grows well except the attack of some pests and put forth one or two new extension flush. Mango plants are supposed to bear fruits 3 - 4 years after planting. By this time, plant attains normal size, a definite framework and becomes physiologically sound to bear the fruits. Therefore, tree management system of developing plants should be followed until the management system is switched over to that for bearing trees. In the first two years, plants need some special care for better growth and establishment. Following care is taken initially in establishing mango orchard to avoid future problems.

Do not Neglect Plant Basins

Proper aeration near root zone and good sanitary conditions in the basins of newly planted mango sapling is a must for keeping the plants healthy and disease free. Proper growth of young plants can be ensured by timely hoeing and weeding of the basin which is considered to be the essential and should always receive first priority. If intercropping is being practiced, the mango plants should not be kept in neglected condition during pre-bearing stage. After harvest of every intercrops, the plant basins should be properly cultivated and essential nutrients should be supplied. It should also be seen that the pest and diseases of inter crop are not causing damage to mango plants. Any such neglect of young orchards proves costly as fruiting is delayed and longevity and productivity of the plants are reduced.

Remove Wild Sprouts and Flower Buds

After planting grafted trees, never allow to grow buds from any other area except from the grafted bud or shoot. If allowed, the growth of the grafted shoot will become weak and may dieback later.

The scion wood for preparation of mango saplings is taken from senile plants. Therefore, the scion stick carries senility and has capacity to produce flowers for one or two years unless juvenility factor of root stock dominates.It has been observed quite often that the mango plants prepared through asexual methods of propagation start flowering and fruit set during the first year of growth or even at the nursery stage. This adversely affects the growth and vigour of plants particularly when such types of plants are allowed to develop fruits. Allowing flowers in the first year not only weakens the plant but also leads for improper framework. Removal of panicle breaks the apical dominance and allows plant to put forth side branches for formation of scaffolds. Therefore, these inflorescences should be nipped off immediately after emergence so that it may not disturb vegetative growth of the newly planted mango saplings.

Protect Plants from Cold and Hot Air

Newly planted mango saplings face first winter from November-February under North Indian conditions. For the successful establishment and growth during winter, it is necessary to protect the newly planted sapling from cold. In North India, some times, winter becomes more severe leading to occurrence of frost. The frost damages the young leaves, gorwing tips and finally the plant may die. Therefore, the protection of mango saplings for at least one winter or in some cases two seasons is very essential. It is advisable to establish mango orchards in such a manner that the rows run diagonally to the prevailing wind direction to avoid creating a tunnel effect. Before planting of mango orchard, it is essential to reserve some place for planting of wind breaks at the border sides of orchard from which hot and high winds and frost are expected. To protect the plants from frost, thatching of individual plant is essential. They can be thatched with any type of dried grass or dried stalk of maize or *bajra* along with paddy straw. The plant is covered from 3 sides leaving a gap towards east to allow entrance of sunlight and air. In other parts of the country where the temperature during winter do not go that lower degree,thatching may not be required. For protection against frosts, the young mango trees should be covered properly well in time and the thatching should be removed only when the danger is over. Some growers remove the covers too early, with the first warming up of temperatures after winter; but the trees are likely to suffer if late frosts occur. Basin irrigation and mulching and spray of water or 1.0 per cent sulphuric acid before anticipated date may be helpful in reducing the extent of loss. Occurrence of hailstorms is a natural phenomenon which causes partial or total loss of mango crop. Likewise, it must be protected in summer to avoid scorching from hot winds and intense heat of the summer months (Loo).

It is necessary to provide support to the plant by staking in the areas where in high winds is more prevalent. The staking is helpful to keep plant stem straight and save the plant from damage by breaking from graft union. When mango grafts have grown taller, the stem may be white washed with lime to protect from sunscald.

Follow Proper Water Management Practices

Young mango plants must be frequently irrigated during the drought spell in rainy season, winter and summer months for at least one year. When the plants are well established in the field, irrigation may be given as per need preferably on the basis of soil moisture status. During vegetative growth period, every effort should be made to maintain sufficient moisture near root zone of the plant. Generally, irrigation at weekly interval during summer and fortnightly interval in winter along with mulching should be done in the first two years of plant establishment. However, the quantity and frequency may be modified depending on the soil type, climatic condition and method of raising orchard. Sandy soils will have to be irrigated more frequently than the clay soils. Irrigation frequency may be adjusted according to rainfall and evapo-transpiration during each season. *In situ* raised mango plants require very less irrigation as compared to transplanted saplings. A mango plant is having a medium tolerance to poor water quality. Water with EC more than 2000 m cm^{-1} or chloride ion content more than 800 mg/l or is harmful. During rainy season, water should not be allowed to stagnate around the plant as it will affect the root activity adversely. Drainage may be provided near root zone by making slope at lower side. In marginally water logged area, planting of mango at raised plateform can also be done.

Apply Balanced Fertilizers

The mango plant needs more nitrogen at young stage to promote tree growth, therefore, in the establishment and growth phase, higher proportion of nitrogenous fertilizer should be given than phsophatic and potassic fertilizers. As the plant grows, the requirement of fertilizer increases and therefore, the doses should be increased annually until the plant attains full growth. For efficient use of applied fertilizer, it is always better to split the annual quantity of fertilizers in 2-3 parts and apply in the plant basin. It is very essential in sandy soils because the loss of nutrient is more under such conditions. Similarly, the loss of fertilizer can be minimized by avoiding its application in the peak rainy season. As a matter of fact, fertilizer in young mango plants should be applied during October and March whereas the manures should be applied during pre monsoon period (June-July). Fertilizers should be placed as a band spread approximately 30 cm from the base of the plants and should be incorporated to the soil mechanically without severely disturbing the roots. The distance of band should increase as the plant grows. Sufficient soil moisture is necessary after application of the fertilizer mixture. Soil incorporation is not necessary if under tree sprinklers are used for irrigation of trees as fertilizers can be effectively incorporated by irrigation.

Keep Orchard weed Free

Weeds should be completely removed from the root zone of mango plants. They not only compete for water and nutrients applied to plants but also shelters mainy insects and diseases. Use of deep cultivation tools are not recommended in young mango orchard as it disturbs the soil and damage active roots. However, alternative methods to suppress / control weeds should be applied. Uses of systemic herbicides and / or mulching with plastics or organic wastes have been found to control weeds satisfactorily without disturbing the roots. Chopping of the weeds is more beneficial than doing clean weeding around the base of plants. Weeds in between plants and between rows should be controlled mechanically, using a tractor mounted or a hand operated grass cutters or by manual slashing. Dry weed can be used as mulches around the trees. Which ever chemical weedicides is used, it is better to use it around the trees. Care should be taken to avoid direct contact of chemicals with the green tissues of plants. Always use chemicals on actively growing weeds. If weeds are dried, mow the weeds before a rain and apply chemicals when weeds are in an active growth stage to maximize the efficiency of chemicals and effective weed control.

Apply Mulch in Plant Basins

Mulching the trees is an important and very useful practice. Mulches are effective in conserving soil moisture, maintaining root zone temperature, controlling weed and soil erosion. Most of the organic materials including farm waste and by products can be used as mulching material but hard to decay materials with high C:N ratios are better. Materials like rice straw, rice husk, dried weeds, saw dust etc. can be used for mulching purposes. Some materials like fresh coir dust may be harmful as the leachates from these materials may interfere with tree growth .When mulches are used, care should be taken to inspect trees regularly for any termite activity to take appropriate control measures. Termite activity may be increased with mulch application. In the dry zone, care should be taken to prevent trees from fire damage during the dry season.

Do Initial Framing

Providing a better frame is the first step to make plant prepared for heavy yields expected under good management. Almost one year after planting, when plants are established, every care should be taken to maximize the tree growth so that the plant can be pruned and trained to provide good framework with well spaced spreading branches and open center canopy. For a proper framework, do not allow branches at the base of the main stem. Such branches should be removed / pinched off in the early stages of growth to leave a standard main trunk without any branch up to a height of 1m. At this height, if 3-4 side branches are there, allow them to grow as limb but if branches are not there head back the main trunk at 1 m height and ensure 4 well spaced shoots in each direction. These will develop into limbs. After 90-100cm, each limb should have 2 primary branches. Each of these primary branches should culminate into 2 secondary and further into 2-4 tertiary branches and 8-10 growing / fruiting terminals. To train branches as spreading limbs, staking of individual branch can be done. By pruning this way, plants can be trained to have an open, well spaced and spreading canopy. The pruning and training process must be a continued for 2-3 years because all these can not be done in a single pruning or two. Therefore, growers must regularly visit the plants and prune them as and when necessary. Pruning should be commenced in well established plants before any anticipated major growth flush.

Grow Inter Crop and Cover Crop

Clean cultivation throughout the year is not desirable in mango plantation. Therefore, vacant space between the trees can be utilized for growing suitable intercrops. The selection of crops should be purely area based and as per need of the farmer. Preferably, the leguminous crops should be preferred. For more detail chapter-7 may be referred.

Maintain Bio Fencing and Wind Break

Bio-fencing should be maintained on sides of orchard to save the young orchard from cattle and wild animal. The method of bio fencing is given in this book in chapter-5 which may be referred for the purpose.

Protect from Insect Pests and Fire

During dry periods, termites may damage underground parts of the tree. To overcome this problem apply a band of engine oil on the base of the stem up to about 45 cm height. Similarly, the measures to control the pests and diseases should be taken in time and time to save the developing mango plants. For more detail refer the chapter on insect pest management.

In the dry zone, heavy winds blow during May-June. Most cultivated lands are damaged by fire at this time of the year. To protect the mango cultivation by spreading fire, have a 5-10 m wide fire belt around the orchard plowed and harrowed to remove dry grass. Also during the dry months of the year, keep an area of about 3 m around the base of trees weed free. This might protect trees even if a fire spread over the orchard.

❑❑❑

9

Ensure that Plant is Nutritionally Balanced

Mango roots uptake nutrient from large volume of soils. Therefore, plant is able to sustain growth even in low fertility soils. Six distinct growth stages i.e. the nursery stage (0-1 year), the field establishment and non-bearing stage (1-3 years), the adult bearing stage (6-10 years), full bearing period (11-40 years), and the ageing stage (40 years onward) are found in the life cycle of a many plant. The nutritional requirements in each of these periods differ. Speedy growth of rootstock and proper establishment of grafts are the chief objectives in nursery,therefore, the nutrients available in the potting medium are the main source of nutrition at this stage. At the time of field establishment, the main objective is to ensure least mortality whereas during the pre-bearing stage the main objective is rapid growth of the tree and the development of a sound and strong frame work which can be supplemented by regular application of manures and fertilizers near root zone. During the full bearing period, good cropping, regular bearing and high quality of fruit are the main criteria whereas, post rejuvenation ability to produce early and prolonged fruiting are the main aim in aged mango orchard. At these stages, plant performance depends on inherent soil nutrient and supplemented nutrition.

Some Points to Consider in Fertilizer Management in Mango

- Mango trees are perennial. They take up nutrients year after year from the soil zone around the roots. If fertilizers are not applied every year, the soil will be poor in nutrients, and the productivity and quality of the trees will be badly affected.

- Nutrient requirement deends on the tree age and its size. More nutrients are needed as the tree develops in age and size.
- Treė vigour also an indicator of extent of nutrition. Weaker trees with less vigor need more nutrients to help them recover.
- Yield recieved during last season also decides the level of nutrients to be applied. The higher the yield in the previous season, the more nutrients are needed.

Role of Major Nutrients in Mango

Nitrogen is needed at the time of vegetative flushing to support leaf growth. In plant system, the nitrogen is utilized for the synthesis of chlorophyll molecules found in green sub cellular organs known as chloroplasts where photosynthesis occurs. In addition to this, N is used as components in many enzymes, co-enzymes, hormones, nucleic acids and many other plant proteins. Nitrogen is highly mobilized and in case of its deficiency, symptoms appear in older leaves leading to development of yellowish color and early senescence.

Phosphorus is required to synthesise ATP, cellular molecules engaged in respiration process to support plant life. It is also found in other bio-molecules such as nucleic acids, enzymes, certain proteins and in cell membrane phospholipids.

Potassium is essential mainly for food translocation within the plant. In mango plant, potassium is generally not found in plant cells as integral components of bio molecules but it is found in cellular sap which facilitates many plant physiological processes such as stomatal opening, activation of enzymes and charge balancing agent in transport of anionic nutrients such as nitrates and sulphates etc.

Nutrition Management

Different doses of manures and fertilizes are required to nourish the mango plant and produce economical yield. Phosphorus stays in the soil a lot longer than potassium. So after the rainy season, potassium is used by the trees in large quantities during the dry season and is easily leached out, therefore, even when enough has been applied to bring the level up, it is recommended additional applications of potassium in the dry season to maintain the level. There are indications that applying phosphorus and potassium stimulates a tree into flowering when it had not yet flowered. Muriate of potash is cheaper than potassium sulphate but muriate of potash is not good for saline soils and if a soil is very low in sulphur, potassium sulphate may be applied. For the convenience of understanding and recommendations based on research and experience gained by the orchardists,the nutritional requirement and management for mango crop are given below.

Nursery Stage

In the nursery stage, the plant requires more nutrition and therefore, the nursery beds / potting mixture should be rich in humus which also improves the water-holding capacity of roots medium and promotes root growth. Normally, 25-40 tonnes of farmyard manure is applied to a hectare where mango stones are sown or the grafts (inarching) are planted after detachment from the trees. When pots / poly bags are used, they are generally filled with a mixture of leaf mould, FYM, top-soil and sand / clay in the ratio of 2:1:1. For improving the growth of seedlings, light doses of inorganic fertilizers, viz., ammonium sulphate and calcium ammonium nitrate, at the rate of 25 to 30 kg of N per hectare is applied once or twice at an interval of about 20 days. While raising mango plants in nursery (rootstock / grafted plants) 3-4 foliar sprays of polyfeed containing 19:19:19 NPK water soluble mixture at 15 days interval has been found effective. If the plants show nutrient deficiency symptom or weak growth, the application liquid manure comprising of cow dung + neem / karanj cake + NPK mixture should be done at monthly interval or on need basis.

Transplanting Stage

It is generally not advisable to apply any fertilizer at the time of planting however, manures, cakes and anti termite inputs should be applied well before planting at the time of pit filling. Almost all soils suitable for planting mangoes, have enough available nutrients to support growth of the trees at least for a year. If the soils are too poor to sustain the plant even for one year, the whole pit soil should be replaced and good soils with nutrient mixture should be arranged for pit filling. In order to meet out the initial nutritional requirement, 25kg FYM + 0.75 kg bone meal / SSP + 2 kg Cake + 50g termiticide may be applied while pit filling.

Once the plant gets established in the field, it requires high amount of calcium, nitrogen and phosphorus for initial growth. Potassium is also required in reasonable amounts. Doses of nutrients vary with soil conditions. In soils with a high calcium content and pH (7.0 to 8.0), surface application of calcium is not necessary but in soils where pH is low, 5.5 to 6.5, it may be necessary to add calcium in the form of dolomite lime. The quantity of lime depends upon the soil reaction and, in certain cases (highly acidic soil) a dose of 2-3 t/ha is given. During the non-bearing period nitrogen is particularly required in sufficient amount to induce vigorous growth. It would be better if some part of N is derived from organic matter so that soil texture, moisture holding capacity and ultimately the growth and development of roots improve. Phosphorus is required to cope up with high respiration rates and for translocation of carbohydrates during non-bearing stage. Potassium is required to meet out the need of photosynthesis but its need during non-bearing period is comparatively less than that of bearing period when it is required for sugar and starch synthesis in greater quantities.

Initial Bearing and Bearing Stages

The mango is a terminal bearer plant. The flower bud differentiation takes place about four to five months before flowering. In the year of heavy flowering and fruiting, the vegetative growth is limited which, in turn, affects the next crop, because the shoots from the earlier flushes in spring season are the only flowering shoots in the next spring season. If the early flush of vegetative shoots do not emerge owing to heavy flowering and fruiting, there will be very few mature shoots available at the time of flower bud differentiation hence, there would be no flowering in the next spring. The manuring of the bearing tree is thus, faced with certain complex problems which the grower must keep in mind in order to adjust his practices fruitfully. In heavy crop years, the objectives of nutrition should be: (i) securing sufficient vegetative growth early in the season for the next year's cropping, (ii) maintenance of tree vigour, and (iii) improving the quality of fruit. These can be partially achieved by application of a heavier dose of nitrogen a little earlier than flowering in the 'on' year to initiate vegetative growth, and / or removal of some panicles at the time of initiation to promote vegetative shoot. Accordingly, the manurial dose for different stages of mango plant has been experimented and suggested to the mango growers.

Some mango trees never produce fruit because the period between growth periods is not long enough to stimulate the growth of inflorescences. In addition, fertilizing mature trees with nitrogen often stimulates the growth of foliage rather than inflorescences. Thus, fertilization of non fruiting trees or matured trees which are about to flower may be a wastage of money.

Table 7: Suggested nutreints/plant/year in different stages of mango

Non bearing stage	Bearing stage	Old trees (above 40)
10-15 kg FYM and cakes 100g N 50gP 50g K	20-25 kg FYM and cake 1kg N 500gP 1kg K	50-60 kg FYM and cake 1kg N 500gP 1kg K

Time of Fertilizer Application

Fertilizers may be applied in two split doses, one half immediately after the harvesting of fruits in June / July and the other half in October in both young and old orchards, followed by irrigation if there are no rains.

Method of Fertilizer Application

First of all, the weeds should be removed from basins. The mixture of recommended dose of fertilizers should be broadcast under the canopy of plant leaving about 50 cm from tree trunk in old trees. The applied fertilizer should be amalgamated well up to the dept of 15 cm soil. To increase fertilizer use efficiency, fertilizers should be applied in 25 cm wide and 20-25 cm deep trenches dug around the tree 2 m away from trunk particularly during June- July.

Foliar Feeding

Foliar sprays of macro and micro nutrients to supplement their soil applications have been found effective to obtain quick response and to correct deficiencies in mango.

Urea can be sprayed singly or in combination with super phosphate and/or potash or growth regulators. The spray of 4 per cent urea increases the length of terminal shoots, number of leaves per shoot, leaf area per shoot, leaf N and water content in Langra and Chausa cultivars and does not cause any injury. Spray of 2 to 4 per cent urea also shortens the period of flowering by 5 to 7 days and improves yield. Thus, urea spray is useful for the reduction in duration of flowering and increase

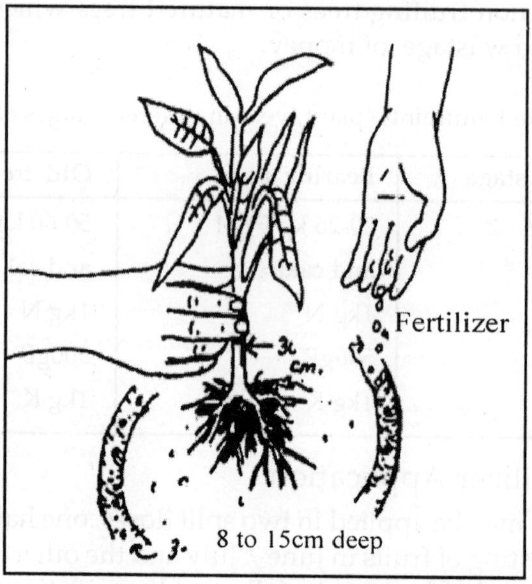

Fig. 12. Fertilizer application method

in number of flowers, fruit set and fruit retention. At pea stage, spray of 4 per cent urea and 40 ppm NAA retains maximum number of fruits per panicle but, at later stage, 40 ppm of 2, 4-D in combination with 4 per cent urea retains the highest number. Fruit weight, sugars, TSS, ascorbic acid, TSS/acid ratio are improved and acidity is reduced. The maximum retention of fruits is obtained by combined spray of 6 per cent urea and 150 ppm NAA.

Micronutrients Deficiency and Correction

Boron deficiency :Boron deficiencies occur on a wide variety of soils, however, boron availability reduces as pH and calcium levels increase. Mangoes are very sensitive to boron deficiency especially when high levels of nitrogen are applied or trees become moisture stressed. Boron is essential for flowering and fruit set, and deficiencies cause flower and fruit drop as well as misshapen fruit. Boron is required for calcium uptake and movement, and calcium deficiencies can be significantly reduced by boron application. Boron plays a similar role to calcium in cell wall stabilization, which makes it essential for quality factors such as skin strength, fruit firmness and storage life of mangoes. Under low boron condition fruits cracking is seen. Lusterless leathery leaves with thickened veins and brown areas in yellow fruit pulp are the other associated symptoms of boron deficiency.

Correction : Application of 100 g borax or boric acid per tree incorporated with recommended dose of manures during July-August or spray of borax @ 4g/l can rectify boron deficiency.

Potassium deficiency : Potassium is an important nutrient for fruit filling (size and weight) and quality. Potassium is required for the production and transport of plant sugars that increase the weight of fruit, while also offsetting the detrimental effects of high plant nitrogen levels on fruit quality. The highest demand for potassium is just prior to harvest. Potassium also promotes more even ripening and enhances skin colour. Because potassium is very mobile within the plant, foliar application of potassium is very effective, especially in the later stages of growth. Soil uptake of potassium is inefficient and therefore, deficiencies often occur in soils high in potash and especially if calcium, magnesium and nitrogen levels are also high. Deficiency symptoms appear in the older leaves that may be a light green colour and have necrosis of the entire leaf margin. Scorching of leaf margins is seen which starts from tip downwards.

Correction : Application of 1 kg muriate of potash or sulphate of potash along with 2-kg urea and 6 kg super phosphate during July-August in the basin could rectify potassium deficiency. Dropped leaves should also be incorporated along with manures to enrich the soil health and fertility

Zinc deficiency : Zinc deficiency is most likely to occur in mangoes grown on alkaline soils, however, deficiencies on heavily leached acid soils may be induced by lime application. Zinc is most important in the early stages of plant development for photosynthesis, root and shoot growth, flowering and fruit set. Deficiency symptoms are normally associated with 'little leaf disease', interveinal chlorosis of young leaves and poor fruit set and yields. Leaves become small and narrow with leaf margins bent upward or downward. Internodal length is reduced drastically and the twig with crowded leaves gives rosette appearance and intervenial areas are pale .Plant is not growing well and the yield, size and quality of the fruit are reduced.

Correction : Zinc deficiency can be rectified by spraying zinc sulphate 2 g/litre water twice at 15 days interval.

Iron deficiency : The leaves loose green colour and turn white and size of the leaf is reduced. The growth of the tree is restricted; yield and quality of the fruit are reduced. In some plants the leaves dry from tip downwards. The symptom described is called "Bleaching". The deficiency is common in soils with high calcium content. Hence, the effect is known as "calcium induced iron chlorosis". Small plants may die due to acute iron deficiency.

Correction: It is best to spray iron chelate as a foliar spray because it is difficult for the tree to take up iron.Two sprays at fifteen days interval with ferrous sulphate 2.5 g per litre of water effectively checks iron deficiency.

Calcium deficiency : Calcium is the most important nutrient for fruit quality and its deficiency is a widespread problem in mango crops where they are grown on heavily leached acid soils which are low in exchangeable calcium and high in available nitrogen. While moisture stress is the major cause of calcium deficiency, high available soil nitrogen or over-use of nitrogen fertilizers that promote excessive plant vigour, also compound calcium deficiencies. Low boron levels also induce calcium deficiency, as boron is required for the maintenance of transpiration (water uptake) and therefore, also the calcium uptake. Calcium deficiencies also occur on soils with high levels of exchangeable calcium and this is largely due to the low mobility of soil calcium and competition with other nutrients such as ammonium nitrogen, potassium and magnesium. For this reason foliar sprays are needed to supplement soil calcium.

Salt Injury : The leaves are scorched and leaves are loosing their natural colour and turn to bronze colour. Tip burning is also seen some plants due to salt injury.

Correction : Raise Dhaincha (Sesbania) as green manure crop with onset of monsoon in the inter-spaces of the orchard during tree bearing years and remove Dhaincha plants with roots at flower bud stage and incorporate in soil of the tree basin. Farm yard manure and compost should be applied adequately every year. Gypsum filled gunny bag if kept in flowing irrigation water,will reduce salt effect.

Copper : If insecticides which contain copper have been sprayed onto the trees, no additional copper should be applied.

Manganese : Sprays such as Mancozeb which contain Manganese fulfil the manganese requirment. Those sprays may not reflect in the soil analysis but are helpful in maintaing the nutritional balance.

Recommended Concentrations of Micro Nutrients for Foliar Sprays

Micronutrients such as Zn, B, Mn, Fe and Cu are essential to all plants. Soils suitable for mango production are generally low in zinc and boron. It is therefore, important to supplement these elements according to leaf analysis. Nutrients can be applied by means of foliar spray once a month after harvest while trees are flushing, at blossom break, at fruit set and once a month after fruit set up to 1 month before harvest. Zinc and boron are compatible and can be sprayed simultaneously, preferably during cooler times of the day. A single spray should be used for minor deficiencies while 2 or 3 sprays should be applied where major deficiencies occur.

Table 8 : Quantity of recommended micronutreint for mango

Deficient Nutrient	Treatment
Boron (B)	Borax at 4 g/ l water
Copper (Cu)	Copper oxychloride at 2 g/l water
Iron (Fe)	Fe-chelate/sulphate 2 g/l water
Manganese (Mn)	$MnSo_4$ at 2 g/ l water
Zinc (Zn)	Zinc sulphate at 2 g/ l or Nitro-zinc at 1.50 ml/ l water

□□□

10

Avoid Moisture Stress and Weed Excess at Critical Stages

Water is a crucial component for mango production. Be it irrigation water or rain water conserved near the root zone for subsequent absorption, water / moisture play an important role in plant establishment growth, fruiting and fruit development.Success of mango orchards depends upon availability of rain or artificial irrigation during critical periods of tree growth and fruit development.The principal objective of irrigation is to apply the required amount of water per tree at right time. This quantity of water should be evenly applied around the root zone which is considered to extend 1 - 1.5 m beyond the canopy in well grown trees. Irrigation systems should be designed at a capacity to provide the maximum water requirement during peak water demand periods of the plant. Water requirements of mango in any stage depends on many factors such as soil type, soil depth, rain fall patterns of the region where orchard is established, plant density, variety, tree size and tree age, etc. All these factors must be taken into account in designing the irrigation systems to increase the water use efficiency and minimize costs of water in case grower has to pay for irrigation water.

During the stage of land preparation, after felling and clearing the jungles, as the case may be, a rough leveling must be done. Care should be taken not to disturb topsoil too much in places where soil depth is low. The degree to which the land is to be cleared and leveled, after all depend on the type of irrigation system that is going to be adopted in the orchard.The following criteria of mango plants are taken into consideration to decide the irrigation requirements.

1. The deep and well spread root system of mango plant is able to collect the water from wide area and depth of soil.
2. Water is essential to solublize the applied fertilizer and its enhanced absorption to put forth vegetative flash and plant growth.
3. Near the time of flower bud differentiation (in terminal shoots of eight to ten months), irrigation leads to vegetative flush which may not mature at right time for flower bud differentiation.
4. The fruit quality is increased if uniform and appropriate moisture content in soil near active rot zone is available during fruit development and maturity.

Factors to be Considered for Optimum Irrigation in Mango

Soil and climatic factors affect the irrigation requirement :The irrigation requirement of mango depends on soil of the orchard and climate of the area. The fine textured deep and well drained soil in Indo-gangetic plains, red soils of Dharwar and the laterite soils which have the property of retention of water, require less irrigation. Very fine textured and sticky soil, like black cotton requires no or little irrigation. Other groups,like light and medium textured soils require adequate surface irrigation, provided they are well drained. Any type of soil having water table up to 2- 3 meters from surface may require no irrigation as the roots of mango tree penetrate to this depth in search of water. Irrigation in mango is also associated with climate including humidity, rainfall and temperature. Mango orchards situated in humid tropics do not require frequent irrigation irrespective of soil type while under dry climate having low humidity and high temperature; irrigation is needed at 15 days interval.

Root depth and its lateral expansion affect the irrigation requirement : Mango tree has deep, well spread and extensive root system except in rocky and hard layered soil. The main root develops secondary roots that go to sub-soil layer. The one set of root system is found in the first 60 to 90 cm of the soil. This extensive type of root system helps the mango tree for its success and in most of the situations. This extensive root system of mango draws and gets the moisture from far and wide. In some odd conditions, the extensive root system development is checked by hard pan rock, stony layer or the dwarfing rootstock used in grafting. In this situation, root zone is restricted and ultimately it restricts the vegetative growth of plant. Trees under such situations, require more frequent irrigation to get good yield and quality fruits.

Stage of plant growth decides irrigation : Irrigation requirement during initial plant establishment, flushing in non bearing plant is essential. In bearing plants, irrigation should restricted during floriferous phase.During bud differentiation, no active growth takes place in mango. Irrigation at this stage may promote vegetative growth under favourable climatic conditions. Therefore, it is suggested that during flower bud differentiation period irrigation of mango plants should be stopped, otherwise irrigation during this period may convert flower bud into vegetative bud which ultimately adversely affects the flowering and fruit yield.

Fruit set in mango remains un affected unless the soil moisture is deficit for longer period. However, soil moisture influences the fruit size, quality as well as the drop of immature fruits. During fruit development period, under hot and dry climate, the irrigation prevents the drop of immature fruits. The moisture deficit in soil brings early maturity to fruits. Regular and normal irrigation to plants during fruit development and maturity period improves the quality of fruits. The fruits are of better size and juicier from irrigated plants than those are under deficit soil moisture. Therefore, regular and timely irrigation of bearing plants becomes necessary.

Precautions/Consideration in Mango Irrigation

Assured irrigation in young orchard

The principal objective of irrigation of young and non-bearing orchards is to get fast and vigorous growth of the plants. In initial stage the root spread of the plants is limited and water capture area is small. Hence, light irrigation at regular intervals is required to wet the soil. The non-bearing trees up to 2-3 years of age may be irrigated at weekly interval. The interval of irrigation depends upon tree age, soil and climate. For the first six months after planting, interval should be 2 to 6 days, for 6 to 12 months old plant it should be at weekly interval and 7 to 20 days till the plants attain bearing age. In light soil, irrigation frequency is more than in heavy soils. During winter, the irrigation is specially required for protection against frost. In heavy soils, frequent irrigation causes damage to root system and stem, so it should be avoided. But the interval should not be too long to create moisture stress which checks the growth and spread.

Judicious Irrigation in Bearing Orchards

To obtain good flowering, the irrigation during winter months (2-3 months) before flower bud differentiation should be stopped as it may promote vegetative growth, which is detrimental to flowering. In North India, 3-5 irrigations are required starting from February (at panicle emergence stage) to May (at full fruit size) at 15 days interval. The maintenance of soil moisture is prime necessity during fruit set and fruit development. It is helpful in attaining full fruit size and reducing fruit drop. In light soils, the interval of irrigation should be short during hot, dry and windy weather than in cold and calm atmosphere. The annual precipitation in most of mango growing regions varies from 100 to 250 cm. therefore, most of these regions require little irrigation. Anticipated irrigation requirement in mango orchard has been indicated in Table 9.

Table 9 : Irrigation need of mango orchard as per as age and growth stage

Age of plant	Growth stage	Critical stage	Irrigation Schedule
1-2	Establishment.	Dry weather condition.	Twice a week.
2-5	Vegetative growth.	Flushing and flush maturity (Nov. - May).	5-6 irrigation.
5-10	Junior bearing stage.	Flushing and fruit development.	1-2 rrigation during flushshing when dry condition and 3-5 irrigation during fruit growth and development.
10 Years onwards	Full bearing stage.	Fruit set to maturity.	5-6 irrigation starting from fruit set to stone hardening and maturity.

Method of Irrigation

A number of irrigation systems like basin, ring, furrow, flood, sprinkler and drip are employed. Each system has advantages and disadvantages as one system may be suitable for one set of conditions but unsuitable for another. Therefore, proper selection of the irrigation method is important for better orchard management practices.

Ring/basin Method : In this system of irrigation, small circular rings are made around the tree trunks in early stage of the plant. The size of rings is enhanhanced as per the age of the plant. Water is applied directly to each plant manually in early stage and through hose pipe in later stage. This mentod need too much man power.

Furrow ring mehtod: This method is similar to ring system where the individual basin is connected by straight channel.The water passing through these channels touch the tree trunk directly.Such type of water flow from plant to plant may cause damage and manures and fertilizers may be washed away with water and deposited at the end of the channel. The disease present in one plant may spread to other plants. In this way, it increases the infection.Therefore, this system is considered as defective one.

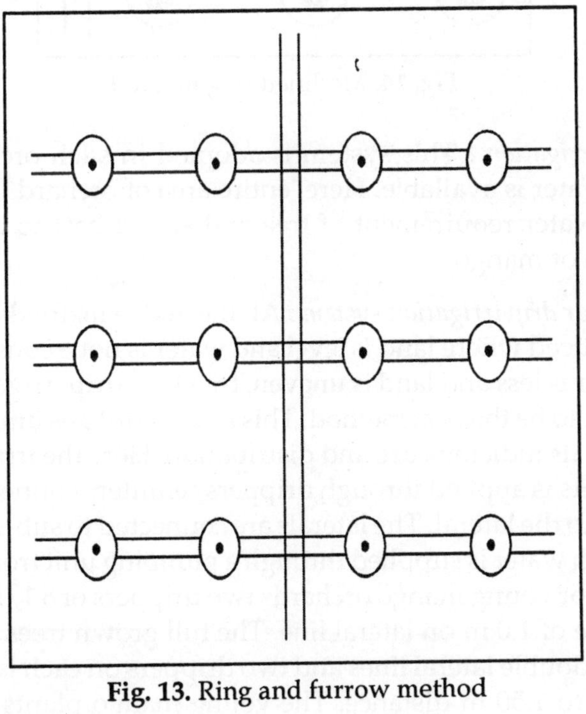

Fig. 13. Ring and furrow method

Modified ring system: In this system of irrigation, rings/basins are made around the tree trunk as per requirement. These rings are connected with a furrow made betwen two rows of mango plant. The water flows in the furrow and one by one each palnt is irrigated through the sub cahnnel attached the ring.

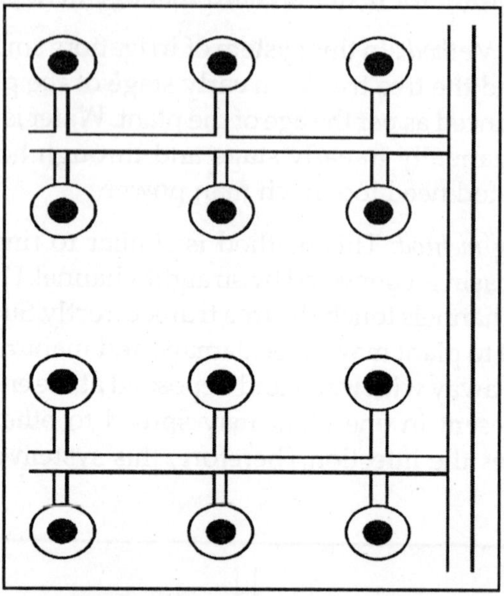

Fig. 14. Modified ring method

Flood Irrigation : This system is adopted in such orchards where sufficient water is available. Here, entire area of orchard is wetted and meets the water requirement of tree and suited best to the extensive root system of mango.

Trickle or drip irrigation system : All the above methods of irrigation can be practiced where land is level and water is not a constraint. Areas where water is less and land is uneven, trickle or drip irrigation method is supposed to be the best method. This system utilizes limited quantity of water for its judicious use and distribution. Here the irrigation to the mango plants is applied through drippers/emitters connected through micro tube to the lateral. The laterals are connected to sub-main or main line in which water is supplied through a pumping unit from the source. Generally, for young mango orchards two drippers of 8 1/ha are placed at a distance of 1.0 m on lateral line. The full grown trees are supplied water with double lateral lines and two drippers on each lateral line are placed at 1 to 1.50 m distance. The young mango plants require 9-12

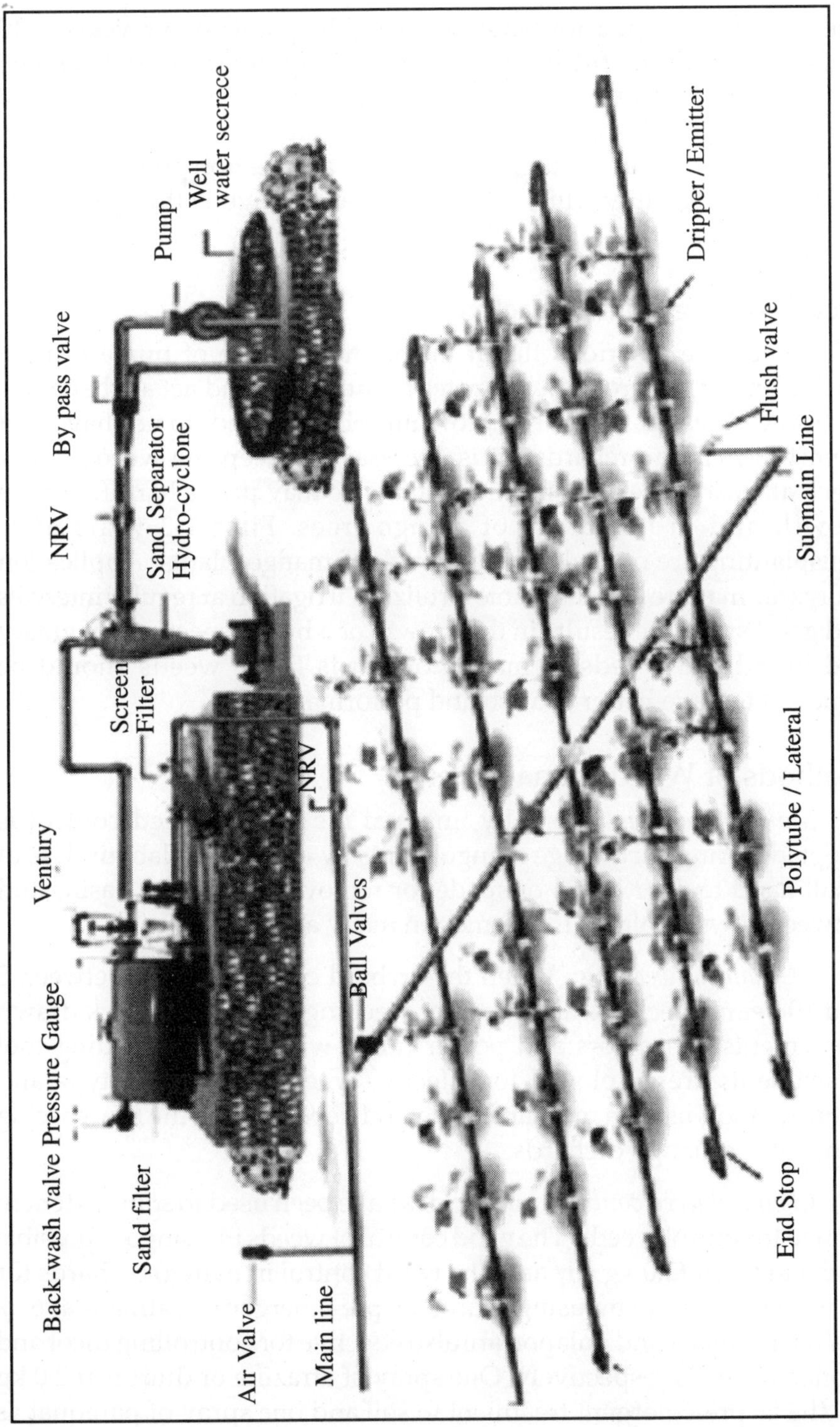

Fig. 15. Animated drip irrigation system (Courtsey, Jain irrigation systems Ltd, India)

litre/day/plant water for better growth. The plants of 3-6 years, 6-10 years, 9-12 years and full grown trees require approximately 30-35 litre, 50-60 litre, 80-90 litre and 120 litre/day/plant, respectively. This system has several advantages such as plant attains quick growth, water saving, weed control, labour saving, easy nutrient application and increase in yield of good quality fruits. The only disadvantage is the high initial cost of installation.

Weed Management Practices

Weeds are a serious threat in the initial stage of nursery plant establishment where they competes for nutrients and act as shelter for pests and diseases, but in full grownup plants, weed do not have that impact.In mango orchards, it is necessary to keep the weeds under check at least twice a year which otherwise may prove harmful to the growth and productivity of mango trees. First 1-3 years after transplanting are crucial for the growth of mango plants. Application of organic manure and inorganic fertilizers, irrigation at regular intervals and good sunshine results in the growth of a broad spectrum of grassy and broad leaf weeds in mango orchards.These weeds should be removed to have better growth and performance.

Methods of Weed Management

Manual weeding : Usually, manual method of weed control is adopted during early stage mango plants by employing labour. Use of small hand tools (*Khurpi*) or spade for removing weeds in basins and between rows of plants is common in many areas.

Mechanical weeding : When the orchard comes to a age between 8 and 10 years, mechanical method of weeding by using bullock drawn implements or tractors and power tillers with special weeding tool attachments are employed for effective weed control. The type and nature of tools used in mechanical method depend upon the row spacing adopted for mango orchards.

Chemical weed control : Herbicides have been used to some extent in mango to control weeds. The good control of weeds in mango is possible with paraquat (3.0 kg a.i/ha). The weed control in mango orchards for 24 weeks can done by using diuron as pre-emergent treatment at 6.67 kg/ha. Bromacil and dalapon are also effective for controlling dicot and monocot weeds respectively. One spray of atrazine or diuron at 2.0 kg a.i./ha as pre-emergent treatment to soil and one spray of paraquat as post-emergent spray on weeds at the rate of 3.0 kg a.i./ha, six months after the soil treatment, is effective for controlling both monocot and dicot weeds for one year in (one to five years old) mango orchards.

The shade effect: In closely planted orchards, where ploughings becomes difficult, the cultivation is done by hoeing with the spade. In more dense orchards (15 to 20 years old), weeds are naturally suppressed due to shade. the degree of weed problem depends upon the variety grown and the row spacing given. There are tall and dwarf varieties and intermediate ones. The canopy arrangement differs in these varieties. When trees are young, the sunlight received at the ground level is more and this results in the growth of perennial grassy weeds. As the tree grows, the canopy cover increases and shades larger area beneath trees which is suppress the growth of weeds. However, in dwarf varieties, more sunshine is received around the trees at later stage also and this encourages the growth of grassy and broad leaf weeds.

Intercropping/Cover Cropping for Weed Management

Cover cropping and intercropping are another practices followed in mango orchards to suppress the growth of weeds, to bring additional income to the grower and to improve the health of trees until the trees begin to bear and can be continued even after that. Generally, cover crops are grown in space between the trees taking due care that no crop comes under the spread of the trees so that roots of the intercrops and the trees do not come into competition with each other for drawing nutrients from the soil. The detail of interropping and cover cropping is presented in chapter-7.

Mulching for Weed Management & Moisture Conservation

The process of covering plant root zone or entire basin with protective coverings is called mulching. Mulching is beneficial to all mango orchards. It is very much useful in plants raised under drought conditions or in adverse soils. The covering materials may be organic or inorganic in nature. Using mulch can help a mango orchard in following ways:

- Mulches make unfavorable conditions for seed germination of weeds and provide a physical barrier for emerging weeds. A good mulch layer saves many hours of manual weeding. Thick layer of mulch material is very effective in preventing the number of annual weeds in the orchard, since they have difficulty to penetrate such a layer.

- Mulches maintain uniform moisture conditions in the orchard. Moisture loss through evaporation is reduced and soil erosion is decreased in rainy season. Slow and steady water infiltration takes place in the soil and crusting is prevented. Mulches reduce the force of raindrops, thereby protecting the structure of the soil at the surface. Mulches may serve as vapor barriers, thus reducing evaporation of soil moisture. Water fall on bare soil surface causes splashing of soil onto the fruit. Mulches keep fruits cleaner and help in reducing the spread of disease.

- Organic mulches add nutrients and humus to the soil as they decompose, improving its tilth and moisture holding capacity. Some mulches are tilled into the soil before planting a new crop and therefore, may have an effect upon soil fertility and soil property. In the short term, mulches may decrease nitrogen available for mango crop. A material that has a high carbon content and is very low in nitrogen and other nutrients may actually "bind" or immobilize plant-available nitrogen temporarily. This occurs because soil microorganisms use available nitrogen to metabolize and decay the organic material. The immobilized organic nitrogen can be made available (mineralized) later as the organic matter continues to decompose.

- The increased organic matter content due to mulches increases soil aggregation, infiltration, and water-holding capacity. In poor soils, addition of organic material improves water retention and reduce water losses through deep percolation.

- Mulches shield the soil surface from solar radiation effects. Because of the increased moisture content and reduction of incoming solar radiation energy, a mulched mango plant basin has lower difference between day and night temperatures. However inorganic mulches can act otherwise. For example, dark mulch can adsorb more solar radiation and may actually increase soil surface temperature; transparent plastic mulches may increase temperature through a "greenhouse" effect; and can prevent water from entering soil, thereby decreasing soil moisture and increasing water runoff.

- Thus, mulches are very much useful for grownup as well as developing mango plants. In bearing plant it has direct impact on fruit drop, fruit cracking and quality development where as in eveloping plants it helps in better survival, growth with less water.

Natural as well as synthetic materials are used for mulching. Details of some of the mulch materials has been given as under.

Different Types of Mulches and Their Use

Organic mulches

Saw wood dust :A layer of sawdust provides good weed control. Five cm thick mulch of saw dust adds nearly 450g of nitrogen if applied 10 cubic feet around the basin. Fresh sawdust contains high carbon and low nitrogen, and its breakdown requires that microorganisms take nitrogen from the soil. It is very good material for mango plants particularly to new plants. It is readily available from sawmills but it tends to be expensive.

Straw : A 15 cm thick layer of paddy/legume straw provides good annual weed control. These materials may decompose quickly and therefore, should be replenished as per need. They stay in place and will improve the soil as they decay. Avoid straw of weeds that is full of viable seed. Legume such as cowpea, supplies nitrogen as it quickly breaks down. Wheat and paddy straw are readily available in villages.

Pine tree needles : Pine needles are also used as mulch in hilly areas. They make excellent mulch for mango but their availability in mango growing areas in plains is restricted.

Field grasses and weeds : The layer of mulch is made gradually using dry grass. A 5cm layer of local grass and weed cuttings provides good weed control and moisture conservation. Layer of green grass/weeds gives off excessive heat and foul smell and decompose slower than other organic material. In limited quantity, green grasses decompose rapidly and provide an extra dose of nitrogen to growing mango plants, as well as making fine humus. Grasses full of viable seed should be avoided.

Dry mango /other plants leaves :After tree pruning, all the twigs should be collected at one place where dry leaves are collected and used as mulch material. A 5-7 cm thick and compact layer of dry mango/other plant leaves, provides good weed control and moisture conservation. Leaves decompose quickly and improve the soil once decomposed. To reduce blowing of dry leaves due to wind blow, the leaves should be surrounded by a basin bund or these may be used only after partial decomposition.

Crop residue and compost : Apart from above organic mulch materials, compost, crop residues, including twigs and branches etc. can be used as organic mulch material in the mango orchard.

Coir pith: Coir pith is easily available in coastal areas near the coconut industries where, it is a matter of concern as its disposal is not done properly and can be seen in heaps. It is a good mulching material however, its leachates may affect the plants.

Inorganic mulches

Black plastic: One layer of black plastic provides excellent weed control. It is relatively slow to decompose, but broken down by sunlight and should be replaced every two years. It increases the soil temperature by which may cause soil temperatures to rise too much in mid-summer, damaging the roots of plants. A good foliage cover can prevents direct absorption of sunlight on mulch surface. It is easy to obtain, but is fairly expensive. A new type of black plastic has recently come onto the market which has a white, reflective side to prevent the overheating problems.

White polythene: White polyethylene can also be used as mulch material in mango orchard but it is very expensive. In increases the soil temperature and also not much effective in weed management hence, not used commercially.

Method of Mulch Application

Most of the organic mulches should be applied after plants are well established. In developing mango plants also organic mulch can be applied but problem of termite should be taken care off. Remove all weeds before spreading the mulch. Apply mulches when there is reasonably good soil moisture and before the weather turns hot. Infiltration of rain water will be slowed somewhat by a mulch, so it is best not to place the mulch on soil which is dry. Water thoroughly or wait for a good soaking rainfall before applying any mulch. Application of organic mulch in plant basin can delay the termite infestation to tree trunk for some days as termite feed on mulch material and do not move to tree. However, it aggravates the infestation in subsequent days when mulch is digested and no additional mulch is replaced. To keep these termites in check ,tree trunk should be pasted with chlorpyriphos 0.02% + coal tar paste and mulch should be applied one feet apart form the tree trunk. This practice delays the termite infestation and restricts the soil mounding on tree trunk. Black plastic and similar materials should be spread on land that has been completely prepared for planting and has a high moisture level. Edge of the mulch should be buried with soil to prevent it from blowing away due to air.

□□□

11

Manage the Canopy and Rejuvenate the Old Trees

Besides the various biotic and abiotic factors responsible for low yield and quality in mango, overcrowding of branches,less circulation of air and light inside the canopy leads to production of more unfruitful and unreachable but fruitful branches in a mango tree. It is mainly due to lack of knowledge about canopy management. In order to make mango competitive both in the domestic and export market, there is a need to use modern production technology i.e. high density orcharding, mechanization and canopy manipulation. Developing desirable plant ideotype is an important tool for mango plantation as it allows the producer to restrain the exuberant vegetative growth of mangoes to manageable sizes and forms. There are no hard and fast rules for pruning rather a matter of common sense; the main objective is to develop a good tree structure which meets desired requirements and facilitates harvesting and movement of machinery through the orchard.

Understand the Importance of Ideal Mango Canopy

- Most of the orchards are dense and over-crowded and their bearing is less. They harbour various pest and disease-causing agents due to change in micro-climate inside the canopy. Pruning is usually carried out to shape trees and open up the centres, allowing free movement of air and sunlight into the tree. This facilitates the penetration of sprays through the trees making control of pests and diseases much more efficient. It reduces the cost of plantation and management by reducing the incidences of disease and pests and also expenses on their management.

- The ability of sunlight to penetrate the tree enhances the colour of the fruit and improves quality. Mango tree bears 90 per cent fruit or one can say produce commercial yield from two third of lower canopy in the peripheral of the tree. Therefore, developing ideal canopy in mango is very much important from the beginning of the plantation.
- It lowers the cost of harvest by reducing the time and required labour for harvest. This is also important to reduce production and harvest costs, synchronize vegetative growth and flowering period, extend the production cycle and sales season, increase productivity of the plants, stimulate precocious flowering, extend the productive life of the plantation, recuperate overgrown in older plantations and increase air circulation inside the plantation, which ultimately lowers costs and losses associated with disease and pest.
- Specific activities such as irrigation, pruning and harvesting at field level and post harvest management operations like grading and packing can be synchorize if all the mango plants have ideal canopy. Related services like pack houses, refrigerated-van, labour force can be efficiently utilized when the plant canopy is ideal and uniform.
- Ideal canopy improves the appearance and quality of mango fruits and helps in predicting yield of a tree which ultimately helps in planning for marketing and export of fruits
- Pruning has direct impact on yield. Mango has tendency to bear at the terminal ends of the shoots, however, some fruits are also produced from auxiliary panicles as well as old branches. The pruning activate the quiescent fruit bearing buds owing to the redistribution of the endogenous hormonal substances, to favour flowering and fruiting.
- Pruning results in, increased dry matter content of leaf and stem which reaches the maximum level by October-November, thus the C/N ratio of the shoot reaches the highest level during November-December. This becomes possible owing to increased

penetration inside trees.

- As a result of pruning, the RNA content of leaf, which attains the maximum level at the time of flower bud formation, becomes higher during the second 'normal' year than in the first 'lean' year. The second peak in ascorbic acid content in leaves during November after the first in September synchronizes with the time of flower bud formation. The pruning increases the ascorbic acid content in mango leaves from September onwards which helps in more flower and formation. In different cultivars of mango, the ascorbic acid content of leaves reaches a major peak in November coinciding with flower bud formation.
- Pruning also increases the total phenolics content in mango. Phenolic acids act as inhibitors similar to abscisic acid which favours the flower bud formation in mango.Lower IAA Oxidase activity has been observed as a result of pruning which possibly makes optimum levels of auxin available to favour flowering.
- The respiration rate of leaves during September and November-December is increased due to pruning which helps in increased flower bud formation and flowering.
- The gibberellins-like substances in the leaves of pruned trees are found to be lower than those of unpruned trees. The lower contents of GA are associated with normal flowering in mango. Similarly, the higher level of cytokinins which favours the flowering in mango is also increased due to pruning. Higher levels of cytokinins induces the inflorescence by its increased flow in the xylem sap which activates the latent buds from their innate dormancy resulting in cauliflorus flowering.
- Harvest period in mango can be manipulated by pruning trees on schedule to stimulate the growth of inflorescences, flowers and fruit so that the fruit is ready for harvest at the desired time. Mango trees can be managed such that the harvest occurs earlier or later than normal to take advantage of higher-price marketing periods. Synchronizing and timing the harvest period of a group of mango trees may also be desirable to maximize the efficiency of marketing activities

Proper Planning for Canopy Management

Canopy management is a skill cum art which needs a lot of planning and knowledge about behaviour of plant after pruning. Therefore, before taking up any cnopy management opertation follwining points needs to be looked into

Understand the Mango Growth Pattern and Fruiting Habits before Pruning

The growth pattern of all the mango varieties is not same. Some are more vigourous (Langara) whereas others are less (Amrapalli). Therefore, pruning depends upon the variety. All pruning is done after a flush (period of rapid growth) and after the new leaves have turned dark green. Mango trees do not have to be pruned annually to bring on flowering or increase yield as is the case with deciduous fruit trees. They are terminal bearers, and usually flowers are borne on mature wood. In fact, pruning at the wrong time may have a detrimental effect. Pruning tends to stimulate shoot development in mango trees usually resulting in vigorous vegetative growth. If trees are cut back too heavily, they may not fruit for two to three years.

Visualize the Image of Desired Plant Canopy to be Developed by Pruning

The crop canopy refers to leaf area and its orientation, height and tree girth. The quality of the canopy is best measured as the percentage of light falling on a field that is intercepted by leaves. Normally leaves should be oriented in such a way that light passing by one is captured by the leaf below. Thus, canopy management is one of the most important steps in orchard management which is done with a goal to shape the tree for maximum light utilization and efficient cultural operations in the coming years.

The ideal canopy for mango may be variable depending upon the needs and requirement. For high density, the small canopy will be ideal where as for wider plantations the canopy should be little bigger. At any cost, the canopy of two plants should not touch to each other. This can be managed by regular pruning. Severe and regular pruning is essentially needed in high density mango plantations which should be done just after crop harvest

The desired plant canopy can be visualized by seeing a perfectly pruned tree and gaining experience in plant canopy by visiting top gardeners, research stations etc. When one becomes able to visualize a desirable tree shape for a mature tree, it will provide significant assistance to pruning decisions on the young tree. Knowing which shoots and branches to leave on young trees to form the basic tree framework is critical, as these shoots will remain throughout the life of the tree. The ideal tree should have three and not more than four trunks, be open inside and low-set, i.e. 4-5 metres. Over this height, harvesting becomes difficult.

Try to Keep a Balance Between Vegetative and Fruitful Growth

A good balance between vegetative and fruitful growth should be maintained by pruning. Factors such as soil type, climate, variety, and nutrition play critical roles, however the severity of pruning is very important. On vigorous varieties, there is need to increase fruit and slow down the vegetative growth. It has been observed the slower vegetative growth usually favours more flowers and increased fruitfulness. Selective pruning to remove vigorous, less fruitful shoots and to retain less vigorous shoots can assist in swinging the balance from vegetative to fruitful production.

Decide Proper Time for Pruning

Generally, the bearing mango plant should be pruned after fruit harvest. The developing plant, however can be pruned during dormant season (November-December). In bearing trees, delayed pruning results in weak growth of sprouts.

Decide Proper Method of Pruning

Marking of the trees is essential before pruning to avoid error. Cut the side branches at least 50 cm from the target. Cut the remaining branch close to the main branch and parallel to it. Make a clean cut to avoid later infection. The cut surface should be painted with Bordeaux paste or clay+cowdung, or coal tar to seal the surface and promote healing of the wound and remove small branches and dead twigs from the upper half of the canopy.

Methods of Framing of Newly Planted Orchard

In the initial years, appropriate frame of the tree should be developed through required pruning. To restrict the tree size in manageable limit, developing an ideal canopy and centre opening with regular pruning becomes essential and useful in early years.Mango being an evergreen plant, requires very little pruning however, the initial training provides good architecture of the plant which ultimately help in production of fruiting terminals. Most of the mango cultivars, grow erect and form a natural dome shape canopy with symmetrical branching which do not need regular pruning. The early stages of growth are most important for the initial shaping of the young tree. A well structured tree will be able to carry a heavy crop, facilitate spraying operations and ensure the fruit is exposed to plenty of light leading to a good blush on the fruit at maturity. The important steps for deveolping a good frame has been illustratred in following text.

1. Generally young trees grow to over 1 metre before they start to push out branches. To develop a strong trunk, the trees should be allowed to grow to over 1 metre in height initially. They are then cut back to a height of between 0.6 to 0.7 m. The site of first cut is important for the development of a strong frame. Mangoes grow in flushes, each flush is delineated by a concentrated whorl of leaves on the stem. This is referred to as a "ring of buds", as a bud capable of forming into a branch is situated at the base of each leaf. Below the "ring of buds" the leaves are more spaced out along the trunk. This is an ideal place to do the first cut. It will ensure that the branches are well spaced out up the trunk.
2. Cut is made above the "ring of buds" which results in regrowth of a cluster of seven or more shoots.These would need to be thinned out to three or four, if a good tree structure is to be attained. After the initial cut, allow the three shoots to grow into branches of over 1 metre length. These sprouts will become the first lateral branches of the tree.
3. Allow the two or three new branch sprouts (extension growth) to grow in one growing period and allow the new leaves to become dark green. Then cut off each branches about a metre in length, which will give a strong secondary branch for supporting the growing tree.

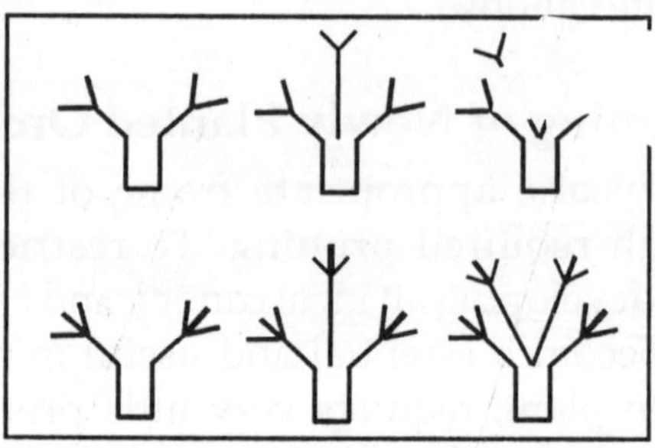

Fig. 16. Different types of framing of mango plant

4. Each of the two or three severed branch sprouts will send out two or three sprouts. These new sprouts should be allowed to grow for one growing period. And, after the new leaves have turned dark green, the growing tips of each of these new sprouts should be cut off.
5. This process of allowing new branch sprouts to grow and then to cut off their growing tips after the leaves have turned dark green is repeated for two years. After this the trees should start branching by themselves.
6. Downward and inward growing branches or branches that cross over each other should also be removed
7. Young trees can be pruned at any time of the year. The aim should be to develop a spreading tree rather than a tall tree. The last tip pruning should be done five months before the farmer wants the tree to flower and nine months before the farmer wants the fruit ready for harvest

Maximizing Production in Bearing Trees Through Pruning

Old mango tree does not produce fruit uniformly. That is, only certain branches of the tree produce during a given season. Often, trees have only a few fruit on some branches whereas most of the tree has no fruit. This means that the tree is not producing up to its potential. A well-managed tree produces fruit on all of its branches. Note that ,if the tree does not produce its maximum, this will not only reduce the return per tree, but will also increase the cost of harvesting as pickers must visit more trees to harvest a given amount of fruit. Pruning all branches of a tree ensures that all branches will produce at the same time. Doing this to all trees in an area or a plantation can bring all the trees into harvest during a relatively short "window" of time and can significantly improve the efficiency of harvesting and marketing activities.

Maximization of production in mango is achieved by pruning, which stimulates all the branches to flower and produce fruit together. It may not be necessary to prune trees every year. It depends on a number of factors. Mango tree growth and flowering follow a known cycle. The length of time between growth periods determines whether the next growth will be foliage (only leaves) or inflorescences that will carry flowers and later fruit. If the period between growth flushes is less than five months, the new growth will usually be foliage. If the period is five months or more, the new growth will likely be inflorescences that will produce fruit. Mango tree normally respond to pruning by sending out a vegetative flush, usually the heavier the pruning the more vigorous and numerous the flushes. To complicate the issue further mango trees are terminal bearers, i.e. they flower from the ends of the branches and will only flower on mature wood. It is therefore, most important to ensure that the trees do not have vegetative flushes near flowering time. Thus, farmers can use this information and knowledge when the trees normally experience a five-month dormant or no-growth period to decide when to prune the tree to stimulate growth that will produce inflorescences rather than foliage.

Times of Pruning

There are two main times when pruning may be done.

First Pruning - After harvest : This should be done immediately after harvest and should be completed by the end of July. Types of pruning at this time include:

Terminal pruning: This includes removal of 10-15 cm top portion of shoots which have produced fruits in the precceding season. This also includes removal of hanging barren flower stalks with 5-10cm branch and leaf.

Skirting : This is the removal of low hanging branches which could hinder orchard operations such as fertilizer application and under tree spraying for weed control.

Centre opening up : This is the removal of branches inside the tree which cross over or clutter up the centre of the tree restricting the penetration of sun light and sprays.

Hygiene : This involves the removal of any diseased or dead ranches in the tree, which could be a source of infection

Second Pruning - Pre flowering : This pruning is generally done from about the middle of Jan.- Feb.and, if the timing is right, it is followed by a floral rather than a vegetative flush. Unfortunately, if it is done too soon or too late after flowering has commenced, the crop may be lost due to vegetative flushing. The time span when this pruning can be safely done is limited to a quite short period about two to four weeks. This includes :

Fig. 17. Centre opening in mango

Skirting : Removal of low hanging branches which could cause fruit to drag on the ground, making it unsaleable.

Opening up : Removal of twigs and branches cluttering up the inside of the tree, as well as opening up the top. This not only facilitates spray penetration for better insect and disease control but also allows light into the tree, improving fruit colour.

Tip pruning : This is particularly useful where the trees had a vegetative flush just prior to flowering. The young flushes are cut back to mature wood; the resulting flush should be floral. This has an added advantage in that it can also be used to reduce tree size, by cutting back two or even more flushes. Second tip pruning is done after harvesting of fruits to remove the dead flower stalks attached to terminal shoot as they prevent new buds to come in next flowering season.

As far as possible, the final pruning should be timed so that the mango fruit are ready for harvest when the market price is highest.

Restructuring Canopy in Big Trees

Pruning the tips of mango branches to maximize and synchronize production is practically impossible on most trees because they are too tall. The pruning technique requires that workers can cut off the tips of all branches on a given tree with a secateur.This is dangerous or even impossible in the case of the old mango tree because the workers cannot reach the branches, even with ladders. Farmers should consider reducing the size of their mature mango trees to a size that will allow workers to reach all branches of the tree either from the ground or with the aid of a ladder that can be easily moved about the tree.

Method

One method of reducing tree size is to cut back only one or two branches each year, or cut back one side only per year. Only those branches that are out of reach from a pruning ladder should be radically pruned. When cutting through larger branches with a saw always do a cut of 20-40 mm into the under side of the branch to be cut, before cutting all the way through from the top. This ensures a clean cut and prevents the branch tearing away as the cut reaches the end. Large upright branches and trunks should be cut on a slight angle to prevent water resting on the cut surface with the potential to cause rotting. Large branches and main trunks should be cut off in sections, rather than in one cut where they could fall onto branches below and damage them. This method will reserve several branches for continued production. This will also imply that the main trunk of the tree will be severed at a height just above those branches reserved for continued production. The tree trunk should be severed at an angle that will encourage rainwater runoff; otherwise, water standing on the severed wood may encourage disease to the tree trunk.

The cut surface of the tree trunk can be painted with Bordeaux paste. The surfaces of large branches newly exposed to the sun by the pruning should be painted with a light coat of cow dung slurry to prevent "sunburn" which could otherwise kill the newly exposed portions of tree branch. After a period of time, the mango tree will re-sprout from the edge of the cut trunk. Of the several sprouts that emerge, the farmer should select five or six for cultivation and should cut away the other sprouts. Once the selected sprouts have stopped growing temporarily after the first growth period and the leaves of the sprout have turned dark green, the tip of the growing sprout should be cut off just as done to newly planted mango sapling. Farmers should stop pruning the sprouts after two years.

Top Working for Rejuvenation - A New Lease of life for Mango

One of the main handicaps that have led to slow pace of evolution of the mango industry in India is the preponderance of seedling trees which are mostly of inferior type seedling plantations of "desi" mango varieties. The fruits of such trees are essentially "sucked out" rather than "eaten". These are still found in the countryside, where no commercial cultivars are grown. Mango is a huge tree: a single tree growing in the open can occupy up to an acre of land. It can bear fruit for 60 years. The yield may vary from a few kilograms to a ton, depending on the age and bearing capacity of the tree.

There are reports that nearly 30-35% seedling mango trees, which had been bearing good crop of high-quality mangoes for juice and pickle, are old and unproductive and waiting for uprooting and sale for wood across the region. Manuring of such orchards is rare. Thus, due to bad sanitation, the trees are generally affected by mango hopper, stem borer, shoot borer, die- back, gummosis, powdery mildew, black tip and mango malformation. Besides, the trees are erratic in their bearing habit and the fruits produced are mostly of inferior quality, fetching little price in the market. Because of their large stature, it is difficult to apply insecticidal and fungicidal sprays. Thus, such plantings are more a liability than an asset.

The commercial plantations of grafted mango trees that have grown old and are not bearing good crop are also being replaced with new plantations. Planting of new orchard may involve a cost of Rs. 80-90 thousand per ha. Uprooting of inferior seedling trees and other low or non-bearing mango trees below 20 years can conveniently be top worked with scion woods of commercial varieties. The process provides fuel wood worth of Rs.40-50 thousand and orchard space may be use for intercrops.

The process of top working or rejuvenation includes removal of old twigs, restoration of new shoots and grafting of desired variety on theses newly emerged shoots. The process may take 2-3 years to be effective.

Method

The process can be started in the month of December -January. Select unproductive orchard, mark the branches, behead them keeping 30cm stubs on 3-6 major limbs. A clean cut should be given to the limbs with a sharp saw to avoid bark splitting. Start beheading from base to the top. Paste cow dung slurry or copper oxichloride paste (dissolve 2 kg of copper sulphate in 15 litres of water; mix 3 kg of quick lime in 2-3 litres of water and then add the remaining 12-13 litres of water; mix the two concentrated solutions) on cut portion. If the old trees are infected with fungi, do not behead the limbs since the stubs may not sprout. Several sprouts will come up on these stubs. Plough the orchard and prepare it for intercrops. Make basins below the plants. Irrigate the plant and apply 1 kg urea per tree. Control stem borer if it has appeared. Remove weeks sprouts, control foliage feeding insect pests and grow summer season inter crop. Take proper care of inter crop and sprouts.

Select the best one or two sprouts on each stub and graft them with the desired cultivar. Remove the rest of the sprouts gently by giving a clean cut close to the stub so that these may not re-sprout. If any of these shoots are left, they will overpower the newly sprouted grafts. This way one will have 6-12 grafted shoots on each tree. These grafts sprout within 20 days of grafting. The shoot above the sprouted graft should be headed back by keeping only one whorl of old leaves in November. These leaves provide food to the newly sprouted graft. Newly sprouted grafts should be covered with rice trash or plastic bags to save the grafts from frost or severe winters in December. The covers may be removed in February-March. While removing the covers, the polythene sheet used for tying the graft union should also be gently removed.

Care should be taken that grafts is not damaged since these cannot tolerate even small pressure. For at least two years, take care that no sprout comes up on the tree limbs other than the grafted shoots. Top-worked trees come into bearing within five years, depending upon the grafted cultivar. The old superior mango trees (Dashehari-Langra, etc) can also be rejuvenated similarly. If the tree is older, it is better not to head back all the main branches simultaneously, to avoid a sudden shock to the tree, which could result in the splitting of the bark of the main stem. The tree can be converted into a commercial variety in stages. In the rejuvenated trees, select 6-12 out-growing sprouts on the 3-6 stubs during June. Remove the rest of the sprouts. These selected shoots grow very fast. During August, the apex of the shoots may be pinched to check the fast growth. This will also help in the development of side branches. The main trunk of this tree should be wrapped with gunny bag during winter to avoid bark splitting.

The land rendered open to sun after top-working can be utilised for growing crops till the trees come into bearing. With rejuvenation, the benefit of an established root system of the trees can be utilised profitably. These trees can bear a good crop for a number of years once more.

Economic Impact

The decision to radically prune a tree should be studied carefully from a financial point of view. Will the expected increase in harvest value more than offset the lost harvest value? Naturally, there will be a significant drop in mango production with the radical pruning to reduce the height of the tree as cutting back large limbs to reduce tree size is always risky with mangoes as one may lose two or more years' production, depending on the amount cut back. Some of the value of this lost production will be recaptured through the sale of the wood cut from the tree. Pruned tree produce several times more mangos than the un-pruned tree because it will have more branches capable of producing fruits. Additionally, some lost harvest value may be recaptured through pruning practices applied to the re-grown tree to manipulate harvest times to capture higher priced markets.

Basic Pruning Tools

Secateurs : Secateurs are handy for tip pruning and cutting out shoots and suckers. They should be kept sharp at all times. Regular cleaning and a drop of oil prevents the blades sticking. A good pair may cost a few Rupees more, but are usually more robust and stay sharp longer. They often have the added advantage of having replaceable blades.

Lopping shears : These are useful for removing quite large branches, up to 50 mm in diameter. There are two types, one has a cross over cutting action where the blade is curved and passes across a cutting plate. This type is prone to splaying if used on very big branches. The other has an anvil type cutting action, the blade is straight and cuts down onto a cutting plate/anvil. This type is more robust and can cut through larger branches by taking two or more cuts

Pruning saw : This tool is capable of cutting through large branches and trunks up to 150 mm in diameter. These have a fairly narrow, slightly curved blade, which enables them to be used in quite restrictive situations such as narrow forks of trees. Some brands are also designed to fit onto a long handle to enable the user to reach higher branches in the tree.

Other Operations that are Done in Orchard for Good Canopy .

Indiscriminate pruning : Unorganized pruning where removal of random growth of unwanted shoots is done by hand or mechanical hedging machines. It results in strong growth of desired shoots. This is done after major pruning to remove unwanted shoots that arise at unwanted places after major pruning.

Selective pruning : An accurate form of hand pruning that usually involves cutting back to a desired lateral or bud(s). Tip pruning, heading back, skirting and flower pruning are all examples of selective pruning.

Headback (heading) : Selective pruning and removal or partial removal of generally stronger upright growth back to old laterals or basal buds. Excessive regular "heading" can cause strong regrowth and watershoots that will need tip pruning or controlled leader pruning during summer.

Tip pruning : In early years light, selective growing-point removal or pruning of generally strong/ medium upright growth back to newly developing laterals on the same branch. Mostly done in older trees that already have developed proper tree shape.

Watershoots (suckers) : Refers to strong vigorous growth anywhere within the tree framework above the graft, usually coming from dormant buds on mature branches. Do not remove this growth completely, as it has enormous fruiting potential; rather, tip prune or headback to 5–8 strong laterals in mid-to-late summer.

Skirting : Pruning of lower branches or "skirt" of the tree, to aid in weed control, surface mulch applications, mini-sprinkler irrigation distribution and to prevent fruit of the lower branches from touching the ground. The best time to do this is at early flowering during cool weather, when trees are not in a strong vegetative-growth phase. A maximum height of 0.5m is recommended. Repeated skirting by selective hand pruning can eventually result in strong, robust branches able to support fruit without touching the ground.

Stumping : Cutting back trees above graft to bare stumps, usually 1–1 .5m above ground. Comfortable chain saw height is best. Best times for stumping in the subtropics mango—late September and October: Immediately paint any exposed branches and stump with good-quality flat white plastic paint to prevent sunburn.

Terminal bud/growth (dominant apical growing point) : The dominant end bud at the end of a vertical branch that has the potential to produce the strongest growth. The strong vertical growth of this bud often suppresses secondary regrowth of laterals and reduces tree complexity and fruiting potential.

□□□

12

Realize Full Potential of Bearing Orchard

To sustain productivity and quality of fruits in bearing mango orchard, appropriate horticultural practices at right time are very important.Management systems for bearing trees are quite different than that adopted for non bearing trees,especially with regard to nutrition and irrigation management. The main objective in managing bearing trees is to promote the tree reserve level for subsequent flowering and fruiting.Depending upon the age of tree, 2-3 stages have been delineated in the bearing mango tree.The plants with 5-10 years age should be nourished and maintained in such manner that a balance between vegetative growth and fruiting is maintained. At this period, trees should be allowed to grow and maintain sufficient vigor for canopy building and fruiting terminals for producing moderate yield.After 10 years of age, mango plants should be maintained for moderate to low vigour or in other words constant vigour and higher food reserves.The constant vigour can be maintained by annual pruning, centre opening and promoting plants to develop shoots which can bear panicles. The tree reserve levels in mango plant can be increased by two ways i.e. by optimizing conditions required for high rate of photosynthesis and by controlling excessive vegetative growth that consume stored reserves unnecessarily. Photosynthesis rate and excessive vegetative growth can be manipulated by controlled nitrogen and water supply to the trees.During bearing stage, the growers has to think its management for long term productivity of trees and to maintain the sustainability of tree production capacity for long period. To ensure this certain steps are needed to be taken in bearing orchards.

Adopt Proper Nutrient Management Practice

Application of plant nutrients in right amount, at correct time and through right method helps plant to take it up efficiently with minimum losses which covers the whole issue of nutrient management. Proper nutrient management has sinificant effects on tree productivity. The main purpose of nutrient management is to keep nutrient levels in a desired range in the tree and to have balance between growth and fruiting of trees as desired by the grower.

The most important aspect of nutrient management in bearing mango plant is its diagnosis and timing of nutrient application in relation to tree phenology or growth cycle.The diagnosis can be got done by analyzing leaf and soil samples of orchard from soil and plant tissue analysis laboratory.It is always advisable to apply different nutrients on the basis of sample report. However, some recommendations have been made for different age group plants (Table - 9) which may vary from one location to other. In bearing mango plants, application of nutrient should be based on the quantity of nutrient removed by harvesting fruits and falling leaves.Excessive or less use of nutrient affects the crop performance. As per principle, the organic manures (FYM, cakes etc.) should be applied in abundance whereas the macro and micro nutrients (N, P, K, Ca, Mg, S, B, Mn, Cn, Zn etc.) should be applied as per need. It is advisable to use agriculture grade micro-nutrients for application through soil and fine / laboratory grade micro nutrient for foliar feeding.

Green leaves (25 kg/plant) and wood ash (10-15 kg/plant) can be applied additionally under poor soil condition. Full dose of organic manures and half dose of the major and full dose of micro nutrients may be applied in May-June after crop harvest and pruning with the onset of monsoon. The remaining dose of P and K fertilizers should be applied during September-October. The additional dose of micro nutrient if required, should be applied through foliar feeding before panicle emergence and after fruit set. The half of nitrogenous fertilizer should be applied during fruit growth and development.

With the exception of nitrogen, the essential nutrients for growth can be applied to the soil at any time without forcing competitive, non productive vegetative flushing. This does not mean to say that some rationalization of fertilizer application is not needed. However, looking at tree nutrient requirements, application timing can be programmed in relation to growth cycle. Now -a- days, nitrogen management in mango orchard is one of the most important and powerful management tool. Knowledge and learning to use the nitrogen effectively is very essential to improve orchard productivity because the application of nitrogen stimulates vegetative flushing and withholding temporarily favors fruitfulness, however, one has to be careful as overall productivity declines, if trees become nitrogen deficient. The timing in nitrogen application is also critical in balancing the tree between vegetative and reproductive growth. Therefore, if possible ¾ or whole dose of required nitrogen should be given just after harvesting to induce growth flush immediately.

Under normal condition,phosphorus may be added during rainy season preferably along with nitrogen after pruning, whereas under rain fed conditions it should be essentially applied in rainy season. It can be very well combined with other fertilizers. Phosphorus is a slow release fertilizer which is very much required in plants to initiate flowering and fruiting process. Mango plant requires potassium during the period from flowering to fruit maturity when the trees are very active in translocating food reserves to maturing fruits. Trees need potassium during flush development also. Therefore, potassium may be applied after harvesting and again during flowering to fruit maturity.

In addition to above major nutrients, other nutrients that are important include Calcium, Magnesium, Sulphur, Zinc, Boron, Iron, Manganese and Copper. Under high tree management, the requirements of these nutrients may be increased as most of these nutrients are removed from the field with harvested fruits. Under such circumstances external applications may be required.

Zinc and boron are very important for quality fruit. For some reasons, zinc seems to be more effective when applied on the ground, rather than as a foliar spray. It is a good idea to apply zinc in lumps in well watered soil. This stops that zinc is being locked up in the soil. Boron works together with calcium, so it is good to apply both at the same time. In fact boron also needs to be applied evenly.Calcium uptake relies on boron. As an added calcium boost for the trees, calcium spray may be applied during fruit set until the fruit is of golf size balls. The smaller the fruit, the more benefit the calcium spray is to the fruit. K, Ca and Mg all compete with each other, so when K and Ca are applied, the level of Mg reduces. It is better to reduce Mg because excess levels of Mg inhibits the uptake of K. Mg does not seem to be important in mango trees. Excess Mg will result in green coloured fruit, even when it is ripe. For fruit quality,all nutrients/elements are important but the calcium is most important.Poor quality fruit is always lacking in calcium and there is never any disadvantage in having more calcium in the soil, as long as the ratios are not imbalance. Requirements of major and micro nutrients may be decided by leaf analysis. Leaves are usually sampled during flowering to assess nutrient requirements. Deficiency of these nutrients can be met by balance fertilization through soil application and foliar feeding at appropriate time. As a matter of fact, micronutreints except zinc should be applied through foliar feeding. The concentrations of micronutrients to be used for foliar applications are presented in the Table - 8. A single spray should be used for minor deficiencies while 2 or 3 sprays should be applied where major deficiencies occur.

Now for a farmer,it is imperative to understand that trees are not over fertilized with any of the major or minor nutrient as excess amounts may cause problems like excessive vegetative growth, low yields, inferior fruit quality and nutrient toxicity.To make the farmer more understanding, a shedule of stage wise nutreint application to a bearing mango plant (Table - 10) can be adopted as such or with certain modification as per need of the location.

Table 10: Fertilizer schedule for beaing mango

Stage	Major nutrients	Micronutrient
After crop harvest	Apply three fourth N and full P and K	Apply zinc 2g / l and boron 4g/l as regular practice and gypsum and lime as per soil test report
Flower bud initiation	Apply the additional dose as per leaf test	Boron 4g/l
Flowering	————	Boron 4g/l
Fruit set and development	Apply potassium sulfate @ 250g/tree. Apply one fourth of does of nitrogen	Magnesium sulfate + Calcium Sulfate 100g each/ tree at marble size. Boron 4g/l, Zinc sulfate 2g/l.
Stone hardening fruit maturity	Potassium sulfate @ . 500g/tree	Magnesium sulfate @ 200 g / tree calcium sulfate 100g/tree, Boron 4/gl.

Placing the Fertilizer at Appropriate Place

In addition to time of nutrient application, the method of fertilizers placement for effective uptake by the plant is also important in nutrient management. For efficient utilization by trees, fertilizers must be equally distributed on the root zone area. In young mango plants, the active roots are scattered 30 cm away from the trunk in a circular fashion (drip line) and in well grown trees this may extend up to about 1.5-2.0m away from the main trunk. After broadcasting equally on the root zone area, fertilizers should be incorporated to the soil by cultivating the top soil. Care should be taken not to damage the roots too much. If thick mulch with under tree sprinklers is available for irrigation such mechanical incorporation is not required as a number of frequent irrigations may slowly incorporate the fertilizers to root zone. Special care should be taken when applying urea as the source of nitogen. Urea must only be applied when the soils are moist. If not, trees must be irrigated immediately after urea application. Manures and fertilizers can also be applied as bands in the root zone of the tree to reduce application costs. A trench of 30cm wide and 15-20cm deep in the drip area should be prepared and the nutrient mixture should be applied in it. After through mixing of the mixture with soil of the trench, it should be covered with side soils and a raised bund is prepared on the outer periphery of the trench. The practice help improve nutrient uptake and reduces the loss particularly during rainy season.

Lime Application may be Necessary

Lime application is an integral component in mango nutrient management. Mangoes prefer a soil pH of around 6.5. At this pH, most soil nutrient elements become available to plants. If the soil is acidic, finely ground lime or dolomite may be applied to adjust the soil pH. These liming materials may be applied as bands along the tree row or just around the trees on the root zone. The application of lime can be combined with fertilizer application in acidic soil. The exact lime requirement at a particular place can be estimated by using soil analysis results however, for moderately acidic soils (5.5-6.5pH) prevailing in high rainfall zones of Eastern India, 5 kg lime per tree should be applied once in 3 years.

Irrigate the Plant as per Recommendation

Like nitrogen, irrigation is another effective and powerful management aspect in mango to improve the tree productivity as well as to control biennial bearing habit. Irrigation is also important in enhancing the fertilizer use efficiency of trees.

While talking about irrigation management, it is necessary at the beginning to understand the critical period of irrigation requirements and required quantities of water in line with the tree growth cycles. In an annual growth cycle, mango has two critical periods of high water demand i.e. one is after harvesting for flushing which continues upto flush maturity and the second during flowering and fruiting which continues upto fruit maturity. Irrigation in the second stage of plant cycle is very important and essential as it is required for the developing fruits which governs the outcome of the orchard. During these periods, the quantities of water demands are very high and is determined by the variety, tree age and tree size, soil type and climatic conditions prevailing on these periods such as rain fall, temperature, humidity and wind velocity. In most parts of the country, the period of flushing and flush maturity is coincides with monsoon and therefore, there is no need of irrigation except in case of drought. The period of flowering and fruiting and fruit development commence during dry periods of the year and therefore, irrigation during fruiting is very essential. As a principle 25mm water per weak is required in root zone of mango plant but the growers can decide the frequency of irrigation and quantity of water depending upon the water availability. Quality of water also plays crucial role on productivity of mango orchard. For good yield, the irrigation water should contain less than 1 000 ppm total salts. The critical time for irrigation is during fruit development and stress at this time can reduce yield.It is essential to provide adequate water from the first set of flower into fruit up to harvesting.

As a general practice, the irrigation must be stopped after the flush mature up to flowering. This is because mango need 2-4 month dormant period before flowering as a pre condition for the development of floral primodia. If excess water is applied before the onset of flower primodia, unnecessary growth flushes may come which affects the subsequent flowering and fruit production. Therefore, during the period from flush maturity to flowering, irrigation may be completely cut off. Under extreme drought conditions, if trees show signs of water deficiency, light irrigation may be given to keep the plants alive and active in photosynthesis. Usually, in areas where suitable climatic conditions are found for commercial mango production, the period from flushing to flowering must have a dry period with very little rains.Main irrigation schedule should start from panicle initiation.The plant require constant and sufficient moisture during mustard, pea, marvel and prematurity stage.Therefore, ensured irrigation should be there.Final irrigation may be given about one month prior to the expected date of harvesting. The time taken from flowering to maturity also may slightly vary according to the climatic conditions and locations. Growers must record observations on flowering and harvesting dates which are very valuable in deciding irrigation timing.

Make attempts to Increase Pollination

Mango flowers are visited by a number of insects including house flies, wasps, wild bees, butterflies, moths, beetles, ants and various bugs seeking the nectar during full bloom. In the event of nectar collection, few of them, mainly flies and wasps transfer the pollen grain from one flower to other and the process of pollination takes place however, a certain amount of self-pollination also occurs in mango. Many of the female flowers are shed or fail to set fruit, or the set fruits are dropped very young stage due to lack of pollination. Therefore, pollination process in mango orchard should be ensured. For this, keep jaggery soaked cotton / jute bag pieces in the orchard at sufficient places to attract house flies. When mango trees are at the flowering stage, sunflower plants should be cultivated on the sides of bunds. Honeybees, which are attracted by the sunflower, also rest on the mango flowers. This facilitates cross-pollination. Heavy rains also wash off pollen and thus, prevent fruit setting. Some cultivars tend to produce a high percentage of small fruits without a fully developed seed because of unfavorable weather during the fruit setting period.

Performing Pruning After Harvest

Mango trees must be pruned after harvesting. The objectives of pruning are to remove dead or diseased wood, to remove additional growth flushes to allow more light penetration into the leaf canopy to enable the fruiting terminals to produce more sprouts which can bear panicles in the next season and also to control tree height to facilitate cultural management practices. Erect branches are less fruitful compared with spreading branches. Crop maintenance and harvesting from upright growing branches is also difficult and beyond the reach of grower and therefore, these must be removed. At the same time lower branches are pruned to about 1 m from ground level to facilitate cultivation practices such as weed control, fertilizer application and irrigation.

A light pruning of main flush is required in dense growing mango varieties like Willard and Lat Sundary which produce 5-6 or more branches per shoot tip at flushing and make the canopy very dense and crowded. Excess shoots should be removed keeping about 3-4 shoots per terminal.

Thinning Out Excess Fruits

A tree can sustain a certain number of fruit depending on its size. If more fruit are set than this, these fruits may drop naturally. In certain instances, however, even after natural fruit drop, a large number of fruit may remain on the tree and due to this only a few fruit will set in the following year. For a commercial cultivation, this is not a good thing to happen. Remove all the excess fruit. This is important to minimize yield variation over the years. Also it is helpful to obtain a higher proportion of well grown high quality fruit every year.

Managing the Tree Basin

Root zone area of the trees must be kept weed free all the time. Manual cultivation is not recommended as it will disturb active roots. Weed killers with heavy mulch may be quite effective in controlling weeds. Mulching materials may be continuously added every year. Mulching along with chemical weed control and under tree sprinkler irrigation has the added advantage of easy tree manipulation for inducing water stress required as a precondition for excellent flowering from flush maturity to flowering. Under these circumstances, the active root zone may be confined to the top 15-25 cm of soil depth and tree manipulation, therefore, become easy with controlled irrigation.

Adopting Pest and Disease Control Practices

Pest and diseases must be effectively controlled to get a good harvest from bearing trees. Pest and disease activity must be closely monitored from flowering to fruit harvest in order to prevent major pest or disease out breaks which might destroy most of the crop if left unnoticed. Major pests include mango hoppers, mango weevil and fruit fly. The diseases of significant importance are powdery mildew, bacterial canker, anthracnose and stem end rot. Timely action is necessary to prevent economic losses from these pest and disease problems.

❑❑❑

13

Make Mango Orchard to Bear Every Year

The profitability of growing mango is influenced by two key factors: productivity, which consists of yield and quality and supply and demand which rule market prices. The terms such as alternate, biennial, intermittent, irregular bearing in mangos are generally coined to describe the tendency of mango trees to bear heavy crops in one year and very little or no crop in the next year under a set of favourable climate. The problem may be rhythmic or abrupt. When the climate of a particular site is not favorable to the appropriate vegetative reproductive balance for that variety to bear a profitable crop, the variety is said to be unfruitful or shy bearer under given set of condition.

Generally, the varieties having a biennial bearing habit, bear a heavy crop in one year which is termed 'on year' while in the next year they have very little or no crop which is termed 'off year'. It is a proven fact that this tendency of 'on' and 'off' year remains so and goes continuously unless the habit is altered by external factors such as diseases or pests, inclement weather at flowering or by growers deliberate attempts by proper tree management. The management of such irregular bearing becomes important as the profitability of orchard depends on the production of reasonable crop every year. Therefore, an attempt has been made to analyze the problem for growers to get an understanding before taking any step to manage this problem. With the far reaching vision and experience of the scientists, it is now established that the biennial bearing is a complex problem where variety, climate, location and tree management practices, etc. play a key role.

Factors Responsible for Irregular Bearing

Selection of irregular bearing variety

It has been observed that varieties with a long growth cycle show natural biennial bearing tendency. Mango plant shows clearly demarcated and very distinct growth patterns during an annual growth cycle. In a normal case, a variety put forth new growth flushes after harvesting and these flushes mature and remain dormant for some period before the initiation of panicle. After panicle initiation the process of flowering, fruit setting, fruit development take place one by one and finally the fruit is harvested at proper maturity stage. After harvesting, again the vegetative flushes emerge in a short time period and in this way the growth cycle of a regular bearing mango variety continues. Time spans for the periods from harvesting to flushing, from flushing to flowering and from flowering to harvesting vary in different mango varieties. Time taken from flowering to harvesting generally varies from 3.5 to 5 months and based on this, the varieties have been classified as early, mid season and late varieties. Usually for a single variety, time taken from flowering to fruit maturity is fairly constant and may slightly vary with climatic conditions experienced during fruit development. Therefore, the early varieties get more time to put forth new shoots and get it matured in principle, and therefore, may bear regularly provided the new shoots are emerging immediately after harvest. But, it is not the case in varities like Bomabay Green, Himsagar, Malda, Lagra etc. Most of the mango varieties however, takes more than 12 months to complete a growth cycle. Therefore, after producing a heavy crop in one year, they have less time period before next flowering to produce the flush and to accumulate reserves for the next crop resulting into poor or no crop in the second year. As a matter of fact, such plants do not get enough time for physiological maturity of the new shoots and produce no panicle or only few panicles in the next year. However, in the third year, again the tree accumulates more reserves to support a heavy crop. In this way, biennial production pattern continue in most of the mango varieties. It is therefore, very difficult to completely overcome this problem. In varieties, where the new shoots do not require long period for physiological maturity, regular bearing can be possible. This problem can be minimized by 1) planting such a variety which can produce flower in comparatively new / less mature shoots. 2) Pruning of previous years fruiting terminals along with left over panicles immediately after harvest and 3) controlled application of N and water at least 2 weeks before harvesting of fruits to induce an early growth flush.

Growth pattern of mango itself may influence the regular bearing

A number of workers have studied the pattern of growth flushes in mango and have reported different periods of primary and extension growths, depending upon the variety and the environmental conditions under which they worked. In fact, the growth in mango takes place in different flushes, which vary in different parts of the country. Most of the workers strongly feel that early initiation and cessation of growth, followed by a definite dormant period, will help the shoots to attain proper physiological maturity, which is essential for flower bud initiation in them. However, there is now enough evidence to show that the growth of mango shoots is purely a varietal feature and that flower bud differentiation in regular bearing varieties is an annual feature. But in biennial bearing varieties, it is governed by 'on' and 'off' year phase of the trees rather than by the time of origin and cessation of growth of shoots. Extension growth on primary shoots is a varietal feature. The shoot depending upon the variety, may stop putting forth extension growth after May or continue till September or onwards. The potentiality of these shoots to form flower buds will depend on the floriferous condition of the tree, which in turn will be determined by the amount of fruit load carried by the tree in the previous year.

Unfavorable climate may cause irregular bearing

Climatic factors are associated with the biennial bearing of mango trees in 2 ways: *(i)* by damaging the crop directly by destroying the fruit bud, blossoms and fruits, or *(ii)* by creating conditions that indirectly affect the production of flower or fruit on the tree adversely. Frost, high temperature accompanied by low humidity and hailstorm fall under the first category, as they directly damage the fruit bud and developing fruits, respectively and thereby reduce the crop considerably. Cloudy weather and rains during blossoming period reduce the crop indirectly by creating favourable conditions for the spread of mango hoppers and of diseases such as powdery mildew and anthracnose. Besides, mango trees in extreme humid places and under milder climatic conditions may not fruit allowing to their increased tendency toward vegetative growth. Even if they flower, owing to excessive humidity, the pollen is never in a suitable condition to be transferred by insects for pollination purpose. At this time, severe attack of mango hopper, blossom blight or early spring frost destroys completely the fruit blossoms which converts an 'on' year into an 'off year condition. The following year, which would have been normally an 'off year becomes an 'on' year, in which an excessive amount of fruit bud is laid down, resulting in heavy bearing. Consequently, the following year is turned into a completely 'off' year and thus, the biennial rhythm is initiated.

Even if a regular variety is grown under rainfed conditions, irregular bearing may be observed, due to variations in climatic conditions. The main factor affecting the bearing process is rain fall. If rains delay after harvesting, there is no shoot emergence in time and shoot remain immature at the time of flower bud differentiation. Under such circumstances, the time period available for trees to have the rest before flowering and accumulate food reserves is lower than that under a normal season resulting into poor crops in the following season. If mango orchards under rainfed conditions face this kind of irregular bearing, then the only option available to minimize it is to provide irrigation facilities.

Nitrogen and carbohydrate reserves may affect fruiting regularity

The seasonal changes in the carbohydrate reserves and nitrogen content of mango shoots and their relationship with flower-bud initiation have been studied by a number of workers. It has been found that higher starch reserve, total carbohydrates and C:N ratio in the shoots favours flower initiation in mango. The available evidence indicates that nitrogen and carbohydrate reserves play an important role in flower-bud initiation. Perhaps, the accumulation of these compounds may create a favourable condition for the synthesis and action of the substances actually responsible for flower induction in plants. A comparative study of carbohydrates, proteins and amino acids in the bark of juvenile and mature mango plants at the time of flower initiation indicated that there was an increased accumulation and metabolism of these constituents in the mature plants compared with the juvenile plants. The inability of juvenile plants to form flower buds under natural condition hence, be attributed to lack of sufficient reserves and their possible hormonal governed redistribution and mobilization. The optimum number of leaves available on a shoot in the 'on' year, play important role in the the growth of fruits to normal size. For the normal development of a mango fruit, it requires 60-90 fully expanded leaves, if the assimilates are to be met from current photosynthesis. As a consequence the developing fruit withdraws large amount of reserve metabolites from vegetative organs during the 'on' year. This might contribute to biennial and erratic bearing in mango.

Irregular bearing may be due to problems in planting site

If mango orchards have been established in very fertile, deep rich soils or in places with high water tables, or in locations with lot of shade etc. irregular bearing is bound to be a problem. Under such conditions, it is difficult to control the vegetative growth of trees. Especially in places where growth checks can not be achieved between flushing and flowering as a result of too much N or too much water, irregular crops may be produced. Under this kind of situation also, the crop load depend on rainfall. However, heavy crops may be possible only when drought periods occur, because it will affect the availability of N and water to trees which is required for growth control for successful flowering and fruiting under such conditions. To control irregular bearing under those circumstances, chemical growth retardants may be quite helpful. Also change of varieties is another possibility.

Hormonal imbalance and physiology of shoots

Auxin appeares to play major role in the induction of flowering in mango. However, if endogenous auxin is the only factor favouring flowering in mango, external application of synthetic auxins would have initiated flowering in the 'off year. The results so far obtained, do not support it. This indicates that factors other than auxins are also involved in the initiation of flowering in mango. Exogenous application of gibberellic acid reduces flower-bud initiation in mango. This inhibitory action is surprising because fruit crops initiate flower buds during long days and gibberellins are active as flower promoter in long-day plants. This suggests that the failure of flowering in an 'off year may be accompanied by a higher level of gibberellins in the shoot tips. Flowering in mango shoot tips is associated with a high level of endogenous cytokinins. Cytokinin concentration in panicle mango is highest 5-10 days after full bloom and decreases rapidly thereafter.The presence of higher level of certain inhibitors similar to abscisic acid, in the shoots of mango trees during flower-bud initiation than that of 'off' year trees indicate that it may also be involved in the initiation of flowering in mango. Since the inhibitor is antagonistic to both GA3 and auxin, it effects cell elongation, and may perhaps check vegetative growths of mango, thereby providing conditions suitable for flower-bud initiation. However, the exact role of the inhibitor in the initiation of flower buds in mango cannot be assessed until more information is gathered about its chemical and biological properties.

Poor tree management may result irregular bearing

Apart from the variety, growing environment and planting site, lack of proper tree management also leads to biennial or irregular bearing problem. If the trees are subjected to a management system based on tree phenological cycle, the irregular bearing may be controlled effectively. The most important management tools available for control of irregular bearing timely pruning, application of nutrients particularly the N and if required, the irrigation. Therefore, special care should be taken to induce new shoots by way of pruning and applying N and water at right time. In addition to the above managements, improper crop protection also leads to irregular bearing or poor crops. If flowers are not protected from pests like mango hoppers or from diseases like powdery mildew and anthracnose, the crop will be very poor in that season but in next season the crop may be little higher due to the buildup of more tree reserves essential for flowering.

Additional nutrient may not help for regular bearing

The earlier notion of some workers that biennial bearing problem is for most part a nutritional one, appears to be wrong in view of the observations made by subsequent workers. The failure of cultural practices or deblossoming for inducing regular cropping in well grown mango trees goes further to prove that the habit is not an acquired one. It appears to be inherent in the tree from the very beginning and becomes more and more operative as the tree grows in age and in that it may be aided indirectly by environmental factors. The regularity of bearing in mango is a cultural problem which can be corrected by influencing nutritive conditions of the tree. Observations at the Horticultural Research Institute, Saharanpur, had also shown that liberal manuring, irrigation and ploughing given to the 20-year-old biennial bearing trees failed to induce regular cropping in them. The investigations pertaining to manuring of the mango trees from the very beginning of bearing phase have not yielded any consistent results so far. Thus, it is evident that while nutritional requirements of the fruit trees must be met for optimum performance of the trees, biennial bearing cannot be overcome by addition of nutrients.

Excessive crop load may cause irregular bearing

Generally, moderate blossoming is one of the chief conditions of annual fruit bearing in fruit trees. The biennial habit can be minimized by undertaking some measures that reduce the number of fruit buds setting into fruits. Investigations carried out in some of the mango varieties have shown interesting relationship between the effect of the total number of fruits harvested on the emergence of new shoots and their subsequent flower bud differentiation.When the number of fruits harvested is optimum, fewer shoots are produced, which in the following year fail to differentiate as flower buds. In the 'off' year when there are no fruits, large number of new shoots are produced which successfully differentiate into flower buds in the following year. Thus, the fruit load on the tree appears to be the main factor governing production of shoots, their flower bud differentiation and ultimately the biennial bearing. Increased crop load seems to have a systematic depressing effect on many manifestations of tree growth. Apparently, in mango the fruiting is an exhausting process. The number of fruits which are retained till harvest, is a varietal feature. However, the total number of fruits that are harvested is important because of their deleterious influence on the production of new shoots and their subsequent fruit-bud differentiation. Therefore, the load of fruits appears to be the main conditioning factor for 'on' or 'off' year in mango. It thus, appears that the limiting factor to the initiation of flower buds is the presence of a crop.

The maximum available (30 leaves) on a single shoot could not support the growth of a single fruit to normal size in the 'on' year. Therefore, the fruit development depended not only on current assimilates but also to a great extent on the reserve. The utilization of reserve metabolites from vegetative organs during the 'on' year could contribute to biennial or erratic bearing. The load of crop in its very initial stages of fruit growth (within a fortnight of full bloom) inhibits the flower-bud formation for the crop of the following year. Non-bearing and bearing units on the same tree are growing under the same environmental conditions with their alternating potentials for flower-bud formation. In the bearing, units of a tree even when new shoots are initiated which attain physiological maturity quite in time, these fail to differentiate flower buds if such a unit carries an optimum load of crop. No definite information is available whether there is a single factor or a complex of factors deficient in such shoots. Besides, the fact that it is an inherent characteristic of a variety (because there are varieties that do flower in the succeeding season despite crop load), biochemical studies indicate that non-differentiating shoots do differ in certain chemical constituents from the differentiating ones. Perhaps, a high level of starch, some auxin like regulators and inhibitors and a low level of gibberellins may be deemed favourable for flowering in shoots.

Management of Irregular Bearing

Mango flowers are formed at the terminal buds of the most recent mature shoots. Most mango varieties flower once a year during winter or spring following a dormant period. Flower initiation is usually triggered by cool nights and dry conditions. The induction of regular, profuse, early and uniform flowering will undoubtedly ensure higher yields and better returns to the grower. In commercial mango plantations, it is desirable to control the vegetative growth and the canopy size to prevent or reduce alternate bearing and to facilitate cultural practices.The following issues relate to flowering and irregular bearing and should be considered to minimise the cost of inconsistent flowering and irregular bearing.

Cultural Management

Deblossoming and thinning of fruits

Flowering can be reduced when tree carbohydrate reserves are diminished as a result of a heavy crop the previous year. Deblossoming is a more severe form of fruit thinning, employed to conserve the reserves of the shoots which could otherwise be depleted later on in the development of fruits. Thus, the deblossomed tree, instead of developing panicles and producing fruits, puts on new vegetative growth in same season, which flowers and fruits in the next year. Deblossomed shoots in mango are tend to bloom in the following year and partial defloration is important in regulating the crop, since excessive fruiting during one year brings about biennial bearing in the mature mango trees. The partial or complete removal of flowers or young fruits in the 'on' year increases flowering the next year, but only slightly. However, the effect of deblossoming on subsequent fruiting habit of the tree seems to be a varietal feature. While the response of

'Dashehari' to deblossoming is a precise one, 'Langra' fails to show the same response. 'Dashehari' tree deblossomed either half or branch-wise yields a fair crop every year, but no such tendency is noticeable in 'Langra'. Perhaps the response to deblossoming is less marked in over-vigorous or vigorous trees. For flower bud formation, deblossoming of a few individual shoots is of no consequence in mango. Even a small deblossomed branch fails to register any difference in its behaviour the following year. However, branch-wise, regulation of crop in 'Dashehari' is quite possible by resorting to deblossoming only once in 10 years. Thus, instead of taking full load of crop in one year, it can be distributed over two years for better profit.

While thinning of fruits has proved beneficial in reducing alternate bearing in a number of fruits (apple, olive etc.), it is hardly possible to practice thinning in mango due to quite a heavy fruit drop in the initial stages of fruit growth. Moreover, fruits may grow up to 'pea' stage even without pollination and fertilization, but drop down afterwards. Thinning, if at all is to be considered, it must be done before the fruitlets attain pea-size. It would be difficult to resort to judicious thinning in mango either by hand or by chemical sprays. Fruit drop continues till the end of April and panicles thinned earlier might not carry any crop at all. The defruiting experiments revealed that the emergence of shoots early in the season does not ensure flowering in such shoots in the following year. The exhaustion is caused by the fruitlets when these are still in pea stage of growth and the tree takes one year to recoup, thus causing biennial bearing in the mango.

Pruning of plants

Pruning is primarily adopted to maintain a proper physiological balance between growth and fruiting. Early pruning after harvest can improve uniformity of shoot growth and subsequent uniformity of flowering, especially after a heavy crop. Mangoes flower from the tips of branches. Pruning to give more terminal branches gives potential for more flowers. Heavy pruning to reduce tree size can be detrimental to yield for several years. Late pruning can also reduce flowering.Tip pruning prior to flowering has shown to improve flowering in tropical areas. This technique involves light pruning trees back to mature wood just prior to flowering. Temperatures below 15°C affect pollen viability, therefore, later flowering in cooler climates can be manipulated by removing early flowers.

As pruning reduces the fruit load in temperate fruits, it minimizes the tendency towards biennial bearing also. In evergreen fruits like mango, pruning is hardly practised. At harvesting, if the fruits are plucked along with the panicle, light pruning is effected automatically and the tree could send forth new growth from the distal lateral buds, if it has the inherent capacity to do so. Generally, the biennial-bearing varieties of mango do not do so, but regular varieties like 'Neelum' send forth new growths immediately after harvest, which differentiate flower buds in the following season. In the biennial varieties, even if some new growth is produced, these are incapable of differentiating fruit buds in the following season.

Controlled irrigation and manuring

To overcome biennial bearing in mango, a number of cultural practices are recommended. These include open planting with wind-break, regular ploughing, liberal manuring at appropriate intervals, ringing and adequate irrigation. By good annual cultural treatments, trees can be kept reasonably healthy, and total failure of crops which usually occurs between 2 heavy croppings can be averted. Although mango trees can be kept reasonably healthy and in better cropping condition by adequate manuring, irrigation and other cultural operations, experimental evidence is still lacking that they can be made to put on optimum crop every year on the entire trees. Adequate nutrition is important after harvest to ensure good growth and carbohydrate accumulation prior to flowering. Too much nitrogen can give excessive vegetative growth, often at the expense of flowering. Boron is important for pollination and fruit development and must be readily available from the soil or applied as a foliar spray prior to bud break. Trees need a growth check or dormant period to initiate flowering. Cooler weather and low soil moisture encourage dormancy. Withholding water from the tree (after the post-harvest flush) for 2-3 months can help trees to enter a dormant phase and improve floral induction. Water stress during flowering can upset flower development and reduce fruit set. Any stress after flowering can increase the number of fruit dropped.

Smudging

The practice of smudging to induce flowers in the mango is reported from the Philippines. Smudging should be done after the monsoon. The heat causes lesser effect than CO_2 that results in the maximum development of buds. It is the heat and not the smoke but gaseous combustion produced from the smudge fire that causes flowering. The best results are obtained when the last growth is well matured and the terminal buds are well formed. Smudging cannot always be depended upon to stimulate flowering. Since it is a cumbersome process, its practical utility in India seems to be limited.

Others

Besides above approaches, flowering and fruit production of mangoes can be practised based on the indigenous knowledge and experience of the farmer in a particular region. Like cutting a cincular ring around the tree trunk to induce stress by disrupting sap flow is reported. Although, ringing as a means to obtain flowering in mango has been suggested by a number of workers,the effect of ringing on flower bud formation is immediate for the next flowering. There remains no residual effect of it and, it has been observed that ringing cannot induce an 'off' year branch to flower in mango.

Use of Chemicals

Plant growth regulators

It is reported that external application of various growth regulators play a very little role in the correction of biennial bearing phenomenon in mango. It is found that a high level of auxin like substance is necessary for flower bud initiation in mango. No flower inhibiting effect of auxin such as NAA and 2,4,5- T has been observed when sprayed at the time of flower-bud initiation even at very high concentrations. They also fail to induce flowering in 'off' year trees by either spraying auxins or MH (an anti-auxin). Spray of ethrel under north Indian conditions has not given consistent results for the control of biennial bearing in mango. The application of chlorflurenol (morphactin IT 3456) at 250 and 500 ppm overcame apical dominance and induced lateral branching.

Use of Paclobutrazol (PBZ)

Flower initiation is very important in mango because it is the first step towards attaining fruit. Trials have clearly shown that while the extent (quantity) of flowering affects yields, time of flower emergence has a significant influence on time of fruit maturity. Early flowering clearly results in early fruit maturity. Unfortunately, our commercial varieties do not flower regularly year after year. Flowering is also staggered, leading to considerable variation in fruit maturity. PBZ is a synthetic plant growth regulator (retardant), which has been used in tree fruit crops to control vegetative growth and to induce flowering. Unlike the other classes of growth retardants that are normally applied as foliar sprays, PBZ is usually applied to the soil because of its low solubility and long residual activity. PBZ through soil drenching at higher concentrations has been observed to reduce vegetative growth and increase flowering, percentage hermaphrodite flowers, fruit set as well as yield and reduced incidence of floral malformation. Some PBZ residues may remain in the fruit. The maximum residue limit of PBZ accepted by the Food and Agriculture Organization of the United Nations (FAO) in stone fruit is 0.05 mg/kg. It is currently available trade names of paclobutrazol are Cultar,Paclotar and Austar. The application of a small amount of paclobutrazol to the soil after emergence of new flush,significantly promotes flowering and fruiting in the following year. Systematic trials have shown the following benefits from the treatment:

- A significant increase in flowering leading to increased yields.
- The early flowering considerably enhances fruit maturity. Treated trees flower three to four weeks early, which reduces the time to fruit maturity by at least two weeks.
- Visually, the fruit develops a better external colour.

Some Important Considerations in the use of Paclobutrazol

Mode of action

Available evidence strongly suggests that flower initiation depends on the presence of an unknown flower promoting factor or factors synthesised in the leaves. At the same time, there are other factors in the shoots which work against the flowering factor or factors. It is believed that a group of plant hormones called gibberellins act as inhibitors to flowering. When paclobutrazol is applied to the soil, it moves up through the roots into the shoots and, due to its anti-gibberellin properties, blocks the synthesis of flowering inhibitors, thereby allowing the flower-promoting factor (s) to work. Paclobutrazol, basically a growth retardant, which retards the GA synthesis in the shoot and develops a requisite ration between Auxin and Gibberellin essential for flowering.

Method of application

The application of paclobutrazol to soil as a drench around the tree trunk (collar drench) is the most effective method, as it ensures proper uptake by the tree. The required quantity is mixed in approximately one litre of water and poured onto the soil around the trunk in a circular band. Soil application of Paclobutrazol (PP[333]) @ 4 - 5 g a.i. per tree in the month of September has been found appropriate to induce flowering.However,the time of application may vary according to time of fruit bud differentiation. It may be applied every year for regular fruiting, particularly in young trees,however, the expert opinion is must for regular application.In dry conditions, a light irrigation is recommended after application. Foliar sprays have been ineffective.

Stage of treatment

The size of trees and condition of bearing of plant at first application of PBZ is important. It has been observed that plants up to 10 years of age tend to be regular bearer,after that the tendency of irregular bearing is seen.Therefore, it would be appropriate if the PBZ application is started after a certain age of the plant.This depends on the age of the trees and the spacing between them. Apart from promoting flowering, paclobutrazol also restricts tree vigour. Trees should therefore,should be allowed to develop a good canopy before treatment commences. In high tree density situations with closer spacing, it is recommended to apply paclobutrazol early when trees are about 3-4 years old. However, when trees are spaced farther apart, say 10 m, early application with paclobutrazol will reduce canopy size and the fruit bearing area. In such a situation, treatment can commence when trees are about ten years old. Following precaution may be teaken while applying the PBZ.

- Do not use on trees of low vigour or under stress.
- Do not use in orchards under planted with any other crops.
- Do not apply to young trees which have recently been interplanted in orchards and which are required to make vigorous growth.

Tree size and canopy influence the dose

Large trees, especially seedling trees, respond more slowly than young, bearing and grafted trees. The dosage required also varies between cultivars. At excessively high dosages, flower and shoot compaction can lead to increased infestation by caterpillars. If such compaction occurs, the dosage of repeat applications should be reduced. Two sprays of potassium nitrate at 4 g/l at ten days intervals, commencing at signs of flower bud burst, were found to minimize panicle compaction. However, compaction is best prevented by using an optimum dosage. If some body is not sure about dosage and trees are ready for the treatment, he should seek expert advice. It is also important to note that tree size and not age is the key factor for determining dosage. Where there is considerable variation in tree size, dosage may have to be varied. If a tree is under stress, it should not be used. Stress can be caused by moisture shortage due to lack of rain, or inadequate irrigation, high temperature, salinity, moisture excess due to water table, flooding or inappropriate irrigation, severe or successive frosts, poor fertility caused by imbalance of major or minor nutrients, diseases, insects, inappropriate herbicide use and mechanical damage.

Tree management in PBZ treated plants

Any treatment of PBZ that leads to increased production should be supported by good management to maintain tree health. This includes nutrition, irrigation, control of pests and diseases, pruning and skirting. It is desirable to prune and skirt trees after harvest and before the treatment. Unhealthy and weak trees should not be treated with paclobutrazol. Treated trees may produce heavier crops and require a higher level of fertiliser application. Ensure that treated trees are not over watered or subjected to drought stress. In some instances, treated trees may set excess fruit and careful attention should be given to thinning operations.

Storage and disposal of PBS packs

Store the PBZ in a closed and well ventilated cool condition in original container. Do not store it for prolonged periods in direct sunlight. Triple or preferably pressure rinse containers before disposal. Do not dispose of undiluted chemicals on site. If recycling, replace cap and return clean containers to recycler or designated collection point. If not recycling, break, crush, or puncture and bury empty containers. Empty containers and product should not be burnt.

Safety directions

Avoid contact with eyes and skin. After use and before eating, drinking or smoking, wash hands, arms and face thoroughly with soap and water. If poisoning occurs, contact a doctor .

Use of Potassium Nitrate

Potassium nitrate is one of the chemicals which have shown some potential in promotion of flowering in 'off' year' mango trees. The KNO_3

seems to have the capacity for breaking dormancy in buds and especially, if the bud is a differentiated one, the flowering becomes obvious. The concentration of KNO_3 required for flowering is about 10 g/l of a high grade (99%), commercially available material. It has been noted that the fruits of good size and quality can be obtained by three sprayings of 3.0 per cent solution of potassium nitrate and 0.6 per cent solution of sodium dihydrogen orthophosphate in combination at frequent intervals of one week on mango crop after panicle emergence. Application of 3 - 4 sprays of potassium nitrate to the trees during the period of flower bud differentiation after the normal crop load in 'on' year crop either alone or in combination with certain growth regulators like TIBA or ethrel has resulted in promoting flowering directly on fruited shoots to the extent of 30-40 per cent which otherwise would have been unfruitful.

□□□

14

Understand the Role of Pests and Pesticide management in Mango Orchard

Pests, diseases and various disorders cause enormous loss to mango through direct and diect invasion on vrious plant parts. Use of chemicals has been one of the conventional mehtods to reduce these losses, however, due to various unwarranted side effects, now-a-days, pest management is relied upon many other options along with pesticides .The integration of all these options is called IPM (Integrated Pest Management). Integrated Pest Management is a strategy to manage pests and diseases on the basis of a systems approach that looks at the whole orchard ecosystem. This includes understanding how the pests and diseases interact with their plant hosts, with the general climatic conditions, plant health and nutrition and with each other. When implementing an IPM system, growers should select ways to reduce overall pest levels in their orchard and ensure that the management of pests and diseases is compatible with their other crop management strategies. It is important that growers realise that IPM system is updated from time to time in response to biological changes that occur in their orchard and new techniques or technologies are introduced as soon as additional relevant information becomes available.

Key Aspects and Concept of IPM

Although the concept of IPM is broad, there are certain aspects that need to be incorporated into a practical program. Mango growers should select those aspects which they feel could be implemented on their farm and keep in mind others which may become feasible at some later time. The key aspects are:

- Accept that IPM programs are not a tool rather it utilize a range or combination of strategies and tools. Farmers should select those that best suit their situation, including the use of chemicals.These tools are discussed and suggested in forth coming chapters agaist differnet pests and diseases.
- There are many pests and diseases that are present in the orchard. All of them do not cause economic damage. A few of them (key pest / disease) may cause damage to that extent where applicaton of control action becomes essential. Identify them and do not apply pesticide just by seeing a few insect in the orchard.Identify other climatic and cultural factors that may reduce pest numbers or damage. There are many natual enemies of insect pests, these are called predators and parasite. Recognize these beneficial insects, which are found in unsprayed orchards and reduce pest and damage levels on cultivated trees.
- Be innovative and have some basic idea about biology and ecology of the pests, their life cycles, the damage they do, their natural enemies and the effects both of the pests on the crop and the crop on pest numbers.
- Visit your orchard even if there is no fruit, exclusively for pest record. Monitor the incidence of pests and beneficial insects to determine the abundance, life cycles and levels of damage.The pests and diseases vary with age and stage of the tree.Therefore, collect information on pest damage in relation to the phenology (growth patterns) of the trees. Read books, take help of experts or decide yourself for the optimum time to apply control methods.
- If possible, keep a pest record register. Record and summarise the monitoring information so that relationships and seasonal trends of pests' damage and beneficial insects can be cognised within the orchard.
- Accept that some pests may be present on tree crops. Pest residue provides food for natural enemies. Low levels of damage should be tolerated and control is initiated only when pest numbers increase (or tare likely to increase) and raise the level of damage.

- There is a list of pest control options.Consider the control options available and select from the list, which is the best suited to your current circumstances and minimise the use of chemicals.
- There are chemicals which have less effect on natural enemies.Therefore,select chemicals which minimise adverse effects on beneficial insects.
- Natural enemies may survive on many other plant/pests species which have no relation with mango. Learn to search for associations such as those between pests and diseases or those between beneficial insects and weeds or flowers. For example, some parasites may require a nectar source, which is available from flowering weeds but not from the mango trees.
- Keep your orchard vigourous. Make attempts to optimise the nutrition of orchard because pest attacks are usually reduced in vigorously growing healthy trees.
- Realise that centre opening ,skirting and hygeine like pruning operations can reduce some pest problems (e.g. mango scale) by altering the density of the tree, but may increase others e.g. mango shoot borer or thrips) by encouraging flush growth.
- Acknowledge that other factors such as cultivars, tree architecture, soil type, water logging, orientation to the sun, localised aspects (slope of hill or on flat land) may also be important in pest and disease management.

Methods of IPM

Look for insects, diseases, plant growth habits and other visible signs in mango trees by inspecting them regularly. This will enable to know crops and the effects of pests and diseases on them. Knowledge and experience is needed to make the right decisions on pest and disease management strategies. An IPM program can incorporate some or all of the following factors that may affect pest and disease incidence in mango crop.

Cultural Practices

Diversity : Orchard that grows with several tree crops in close proximity often have fewer pest and disease problems than those where monoculture dominates. The pest and disease complexes on the various crops may be different and frequently the growth stages of the different trees are not synchronised. Natural enemies may build up and survive on crops that are not treated and invade those trees where high populations of pests occur. Intercrop/ cover crops provide good crop diversity.

Nutrition : Balanced nutrition is very important. As part of this strategy, growers should aim to keep the nutrition of their trees at an optimum level. Pest attack is usually less in vigorously growing healthy trees. Many insects are attracted to high nitrogen content in leaves whereas high potassium levels have a repellent effect or are possibly toxic to some pests. Application of need based micronutrient helps in maintaining of good orchard health.

Pruning : Opening the tree to improve air movement, light penetration and plant protection chemical to access and remove sources of infestation. Skirting helps in reducing the crawling up mealy bugs and hygiene pruning helps in reduction of diseases inoculums.

Tree shape and size : The shape and density of the tree could affect both the level of disease incidence or pest attack and the efficiency of control measures. Tall, dense and compact tree do not receive sunlight at inner side causing development of many diseases and pests. Such trees do not permit proper application of plant protection measures.

Cultivars : Varieties and cultivars may vary in their susceptibility to pest and disease infestation. Varieties known for susceptibility of particular diseases or insect pest, therefore, should be avoided.

Resistant scion material : Scion should be selected from varieties that are seen to be less susceptible to insects or disease.However, less efforts have been made on this aspect.

*Availability of alternate hosts :*Many fruit trees and forest plants serve as host for mealy bugs and scales of mango. Thus,weeds and other plants may allow pests and diseases to increase within, or in close proximity to the orchard. However, they may also provide food and shelter for natural enemies.

Location of the orchard : Soil type, water logging, orientation to the sun and localised aspects (e.g. slope of hill) may also be utilised to reduce pest and disease incidence on trees within a block. For example, mango scab is more prevalent in low lying areas where fogs may occur during flowering. Similarly mango hoppers are more prevalent where soil is damp and low lying.

Pest free planting material : Bring the planting materials from stock which is free from pest and diseases. Scales, malformation and many other problems can be avoided by using disease free planting materials.

Planting density : High density planting is more susceptible to hoppers and scales in comparison to low density planting. Spraying of chemicals in unpruned dense HDP orchard is very difficult which helps pests to escape the pesticides.

Field practices : Ploughing to expose hibernating pests, regulation of moisture through irrigation, application alkathene band to prevent mealy bugs are some of the practices that prevent pest incidence.

Biological Factors

Maximization of natural biological control : A wide range of natural enemies attack pests of mango. These include predators, parasites and insect diseases. Growers should learn to recognise these organisms and attempt to attract them into their orchards and give them the best chance of survival there.

Application of bio control agents : If necessary, biological agents can be introduced when pest populations are low and released to coincide with susceptible life cycle stages of these pests. Some of these may be purchased from commercial sources or collected and introduced from neighboring areas.

Spot application and judicious use of pesticide : It is vitally important that orchard managers use pesticides judiciously to reduce the effect on the natural enemies of the main pest species. Spot application of fruit fly bait and trunk application of pesticide for the control of stone weevil and hoppers are some of the examples.

Chemical Factors

Pesticides : The use of chemicals is still very much a component of the IPM control strategies and growers should regard pesticides as another tool in the IPM arsenal. Equally, it is important that it is not the only method considered or used to manage pests and diseases in mango crop.

Selective chemicals and bio-pesticides : Growers should choose pesticides that have the least effect on non- target insects. For example, neem products,pongamia oil,virus based pesticide, fungal insectcide, *Bacillus thuringiensis*, selective chemicals and the "softer" insecticides like potassium soaps should be used whenever possible.Home made preparatons such as garlic extract and bio-dynamic products may also be useful under certain situations.

Timing : The timing of sprays can affect the efficacy of the pesticide applications and the presence of pests. Considerations may include the time of day when treatments are applied, the prevailing weather conditions and the phenology of the trees (e.g. whether they are flushing).

Application techniques: Whenever possible, the use of spot spraying rather than total block coverage, should be considered since it will have an effect on both the cost and on the natural enemies within the block. In some situations, only a few trees around the boundary may be infested and require treatment while in others, only parts of the tree (e.g. the trunk) may need to be sprayed. High volume applications and higher pump pressures to achieve penetration of the chemicals within the canopy may also be important.

Others :Use of pheromone and early harvesting have been found successful in management of fruit flies.

Proper Care and Handling of Pesticides

In integrated pest management, pesticides have a major share, though the use of other control methods like biological control agents and bio-pesticides is on the increase. Synthetic pesticides are popular because of the ease of application, effectiveness and high economic returns. Most pesticide accidents result from careless practices or lack of knowledge about safe handling of pesticides. Being poisonous in nature and harmful to human beings and other creatures, pesticides should be used with care.

At the Time of Purchase

At the time of purchase, select the pesticide according to the recommendations of competent authorities. According to the Insecticides Act, 1968, certain information is mandatory to be given on the label of the container. These include technical and trade names of the pesticide, composition (percentage of active ingredient and other auxiliary materials), manufacturer's address, registered uses, designation of the toxicity class, net contents, date of manufacturing and expiry, etc. Purchase the insecticide by trade name of high standard. Do not purchase pesticides after their expiry date.

Before Use

After the purchase, read the label information carefully. The toxicity label is in a diamond shape, divided into two triangles. The lower triangle is red/ yellow/ blue/ green; these colours depict the toxicity class, red being extremely toxic, followed by yellow, blue and green. The handling of the red-label pesticides needs utmost care. It is necessary to ensure that employees are adequately informed about the potential physical and health hazards of working with pesticides and the adverse affect pesticides can have on the environment if used improperly. This is applicable to all personnel who apply, transport or store pesticides. Give proper instructions to pesticide applicators and enable them to learn safe pesticide management procedures. Taking time to use pesticides properly is an investment in the health and safety of the employee, his or her family and others.

Signal Word

To alert pesticide users to the toxicity of a pesticide, a signal word will appear on the label. Signal words will tell the user whether the chemical is highly toxic, moderately toxic, or slightly toxic. There are three different signal words used on pesticide labels. Look for the signal word, prominently placed on the front panel of the label. The signal word will indicate what degree of hazard can be expected when handling the product.

- CAUTION means the product is relatively low in toxicity to the handler or applicator.
- WARNING means the product is moderately toxic to the handler or the applicator.
- DANGER, usually appears with the word POISON and a pictogram of the skull and cross bones, means the product is highly toxic to the handler or applicator.

During Use

While opening the packing, take care not to blow or splash the contents. Use only the recommended dose as a higher dose, besides being costly, can also cause phytotoxicity, residues and other health hazards. On the other hand, a lower dose will not be effective.Mix the pesticide in a ventilated place to avoid inhalation. Do not mix in containers meant for general use. For mixing, use rod and never bare hands. Do not use mouth to siphon pesticide out of a container. Those handling pesticides should not be sick or have any bruise or cut. Wear protective clothing while using pesticides. Do not eat, smoke or chew during the spraying operation and never rub the mouth or eyes with your hands while spraying. One should not work alone while handling the extremely toxic pesticides.Do not spray against the wind, as it will cause drift, leading to harmful effect to non-target organisms, and the wastage of pesticide. Never apply granules with bare hands as these formulations are highly toxic and are absorbed through the skin. Generally, with packs containing granules, gloves are provided; if not, then wear synthetic gloves and use a granule applicator. The risk of skin absorption is increased in warm weather when sweating. Also, people tend to avoid protective clothing in such weather. Do not use your mouth to blow open clogged nozzles, instead use a fine wire.When there is bee activity in the field, avoid pesticide tapplication. But if necessary, then do it late in the evening when there is no bee activity. Most ready-to-use formulations are inflammable; do not bring these near a flame.

After Use

After use, clean the spray equipment thoroughly with water. The unused pesticide should be kept in a labelled container. Do not use empty pesticide containers for any purpose, particularly for storing water or feed for animals. The containers should be destroyed and buried. Wash hands thoroughly with soap, take a bath, and change your clothes. After application of the pesticide, observe the waiting periods recommended for different pesticides before harvest.

Storage and Disposal

Pesticides are to be stored in a location which is dry, adequately ventilated, and is free from fire hazards. The pesticides are to be stored in their original containers that are labelled according to content and the containers are to be free from leaks. The storage site must have an immediate supply of clean water available in case of an emergency. Pesticides are never to be disposed of in dumpsters, sinks, or drains of any kind. Store the pesticides in labelled containers out of the reach of children.

Routes of Entry of Pesticide

Pesticides can enter the body in three major ways: 1) through the mouth (orally); 2) through the skin and eyes (dermally); and 3) through the lungs (by inhalation). People may be poisoned without realizing the seriousness of the exposure especially if pesticides enter through the skin and lungs. Acute exposure is a single incident of exposure to a pesticide. Usually the symptoms of a poisoning begin quickly and leave little doubt about the cause of the illness.

Chronic exposure is repeated exposure to pesticides over a period of time. Chronic exposure may go unnoticed since some pesticides may persist in the body for a long time without any obvious signs or symptoms of poisoning. If a person using the pesticide continues to be exposed to these pesticides, residues in their body may increase. An additional risk is that even low-level chronic exposure may lead to serious illness.

Personal Protective Equipment (PPE)

Most of the pest applicators in India are illiterate and unfortunatly, they are not given proper training and protective equipments by their literate owners.To prevent pesticides from entering the body, the person applying pesticides are to wear personal protective equipment. The lack of any statement or the mention of only one piece of equipment does not rule out the need for additional protection. No safety recommendations can cover all situations. Knowledge of pesticide toxicity will allow employees to assess potential hazards and guide them in selecting the kind of protection needed. As with all types of PPE, it is important to inspect the equipment before and after each use.

Body covering :Any time pesticides are handled, the user is to wear at least a long-sleeved shirt and long-legged trousers, or a cover all-type garment. Coverall-type garments are to be made of woven or laminated fabric, manufactured specifically for protecting skin exposures to chemicals. When handling pesticide concentrates or very toxic materials, a liquid-proof apron is also to be worn.

Gloves :When handling concentrated or highly toxic pesticides, wear gloves. For liquid formulations, liquid-proof neoprene gloves are best. Gloves are to be long enough to protect the wrists. When working with hands and arms overhead, put the gloves outside of the sleeves.

Hat :Wear something to protect the head. A wide-brimmed hat will help keep pesticides off the neck, eyes, mouth, and face.

*Shoes and boots:*While sturdy shoes and socks are sufficient for some pesticide applications, neoprene or rubber boots are a wise precaution.

Many canvas, cloth, and leather shoes can readily absorb pesticides. Wear unlined boots with trouser legs outside the boots so the pesticide will not run down the leg and collect in the boot.

Goggles or face shield :Wear liquid/chemical splash goggles or a face shield when there is any chance of getting pesticide in the eyes. Eyes readily absorb pesticides and may cause temporary blindness. Users may wear goggles alone or with a respirator.

□□□

15

Adopt Preventive Practices and Curative Options for Insect Pests Management

The pest scenario in mango during last one decade has changed because of increase in area and change in agronomic practices like close planting, drip irrigation, regular pruning etc. As a result, several insects are attaining major pest status. Like most fruit tree, crops mango is usually attacked by two or three key pests, several secondary pests and a large number of occasional pests in localized areas. Nearly 260 species of insects and mites have been recorded as minor and major pests of mango around the world. Among these, 87 are fruit feeders, 127 are foliage feeders, 36 feed on the inflorescence, 33 inhabit buds, and 25 feed on branches and the trunk. Out of these, only 30 insect pests have been found to cause loss to crop growth and yield in Indian subcontinent. Fruit flies, stone weevils, tree borers and leafhoppers are key pests and require annual control measures. Secondary pests generally occur at sub-economic levels, but can become serious pests as a result of changes in cultural practices and mango cultivars or because of indiscriminate use of insecticides against key pests. Scale insects become serious pests following non judicious use of insecticides against fruit flies. Similarly, mites are secondary pests of mango, which can become serious because of human intervention. Occasional or incidental pests also can cause economic damage only in localized areas at certain times; the great majority of pests fall within this last category.

Pests of Mango Fruit and Their Management

Fruit flies, stone weevil, fruit sucking moth and fruit borer are the pests that attack fruit of mango. With increased awareness on quality fruit for local consumption and export, insects that damage mango fruit by feeding, scratching or ovipositing in the pulp or seed have become more important. Only fruit flies, seed weevils and lepidopterous larvae actually penetrate the fruit pulp and seed. The feeding of other pests, such as fruit sucking moths often extends only into the pulp of ripening mangoes.

Fruit Flies *(Bactrocera species)*

The fruit flies derive their name from close association with fruits. They are the serious and important pests of mango in the entire world. In addition to heavy damage, they also restrict the import and export of fruits. As fruit flies have global presence, few insects have a greater impact on international marketing and world trade in agricultural produce than fruit flies. Most mango producing countries are located in fruitfly infested areas, and growers suffer significant direct and indirect economic losses resulting from fruit fly damage.

Fruit flies inflict damage to crops in three ways 1) oviposition in fruit and soft tissues of vegetative parts of certain plants, 2) feeding by the larvae inside the fruit and 3) larvae tunnel into the fruit, contaminating and providing entry for fungi and bacteria leading to decomposition of plant tissue. The female fly has tapering abdomen which ends in an ovipositor . After mating, females search for a suitable fruit and lays eggs.After hatching larvae feed in fruits which is the most damaging. Larvae feed on the pulp of fruit which appears normal from outside, leading to rotting and becomes unfit for human consumptions. Infested young fruits become distorted, callused and usually drop. Mature attacked fruits develop a water soaked appearance. Attacked fruits usually show signs of oviposition punctures; ripe fruits have high sugar content and exude a sugary liquid. Even when only a few eggs are laid, fruits become ugly in appearance and have reduced marketability due to egg laying punctures and internal decay. Fully grown larvae drop to the ground and enter the soil, where they pupate. The egg, larval and pupal period depends on species and climate. Extent of damage by fruit flies depends upon type, size and condition of fruit, availability of different hosts and population density.

Management

Bait Application Technique (BAT)

Adults (both male and female), after emerging from pupae, require protein and sugar for maturation.They can feed on jaggery, molasses, protein hydrolysate, yeast, fruit pulp (banana) and sugars etc. Taking the advantage of this behaviour of fruit flies, the poison bait (sugar or protein and insecticide in one formulation) are prepared and placed / sprayed in the orchard. The technique of using food bait to control fruit flies is commonly called as BAT (Bait Application Technique). Bait sprays work on the principle that fruit flies are strongly attracted to a food source from which ammonia emanates, and ingest a lethal dose of insecticide together with the bait. Bait sprays have the advantage over insecticide sprays in that they can be applied as a spot treatment so that the flies are attracted to the insecticide and there is minimal impact on natural enemies. Attraction to the bait allows the use of spot sprays, rather than broadcast applications of insecticides. 'Hot spots' of the fly should be located to apply bait sprays.

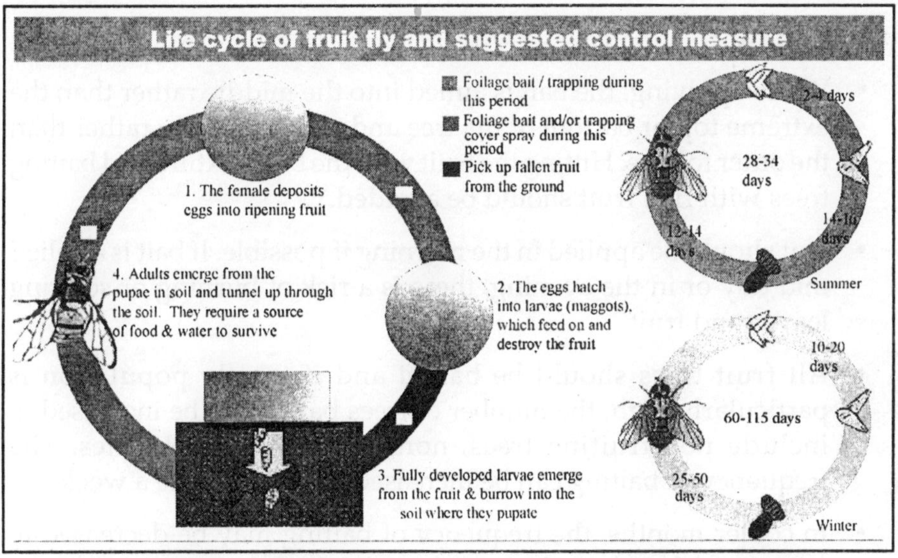

Fig. 18. Life cycle of fruit fly and sugggested control measure

The baits are eaten by the adult fly, which forage over the leaves for food which kills a large number of flies. Due to large scale killing of adults of fruit fly, the number of egg lying female is reduced and infestation goes down. BAT is done in two ways; spraying of bait on trees and foliage or its placement / hanging in pots at different spots.

Food bait for placement can be prepared by mixing banana or gur 50 gm in earthen pots or bowls supplied with 5 gm of borax and 5 drops of insecticide. These can be placed at 10m distance. Once bait is rotten or exhausted and don't attract the flies, these are replaced with new batches of bait. Spot spray of bait is more effective than its placement in pots.

Spot Application of BAT

- Prepare liquid of 0.1% insecticide and 10% jaggery or 10% pulped ripe banana in water. Apply in spots of 40 ml at a rate of 200 spots/Ha (approximately 8L/Ha)
- Apply with sprayers or with a fine brush or broom to the undersides of leaves, about 1.5m above the ground in orchard to coat leaf surfaces smoothly.
- BAT mixture may be of either jaggery or banana or a mixture of both and insecticide.

Precaution

- While applying, the bait is aimed into the middle rather than the extreme top or bottom of the tree and into the centre rather than the outer foliage. Hitting the fruit with the bait mixture and baiting trees with ripe fruit should be avoided.
- Bait should be applied in the morning if possible. If bait is applied mid day or in the evening, there is a risk of burning or scarring leaves and fruit.
- All fruit trees should be baited and if the fly population is particularly high, the number of trees baited can be increased to include non-fruiting trees, non-fruit trees and bushes. The frequency of baiting can be increased to at least once a week.
- In cooler months, the frequency of baiting may be decreased to once every fortnight. Rain washes off foliage baits, so after heavy rainfall reapplication of bait is needed.

Male Annihilation Technique (MAT)

The male annihilation technique is the most suitable method available to date for the eradication and control of orchard flies. It involves the use of a high density of bait stations consisting of a male lure combined with an insecticide (usually malathion), to reduce the male population of fruit flies to such a low level that mating does not occur. Mehyl eugenol (sex parapheromone) is the most powerful male lure and quite long lasting under suitable conditions and attracts male of orchard fruit flies. However, concern over the carcinogenicity of methyl eugenol calls for the development of other types of parapheromones to attract *Bactrocera* fruit flies.

Method

- Prepare 5 x 5 cm squares of plywood of approximately 1.2cm thickness. If ply wood is not available mango wood of same size can be used.
- Prepare a mixture of ethanol: methyl eugenol:malathion or DDVP in the ratio of 6:4:1 V:V. Soak the ply wood blocks in solution for 48 hours. Normally one block soaks nearly 13-14 ml of above mixture (7.60ml ethanol +5 ml methyl eugenol +1.27 ml insecticide).
- Nail the soaked blocks to plant stem or hang them in trees at a density of 10/Ha at about 1.5m above the ground below the branches .
- In steady and prolonged rain soaked blocks are wetted, deteriorated and loose their potency. Protect soaked blocks from

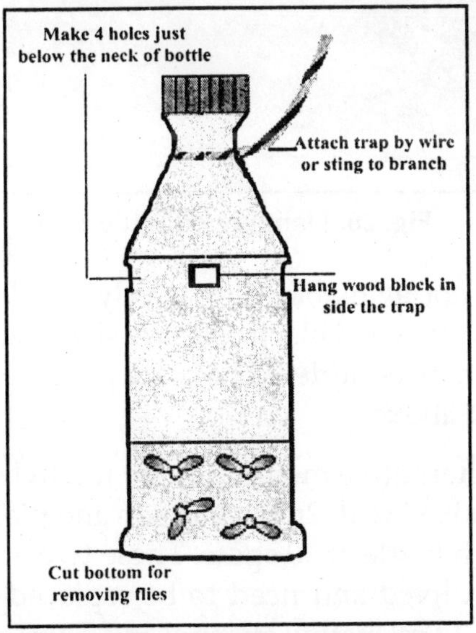

Fig. 19. Preparation of bottle trap

the rain by putting them in traps. In market various traps are available, however, to avoid the cost, a home made water bottle trap can be prepared and used.

- Once prepared, blocks have short shelf lives (2-3 months) and fresh blocks should be made for each deployment.
- MAT blocks do not need to be in traps to work and can be hung using string or nailed in the plant stem, though, these often give information and reassurance to farmers, if placed in a trap. Catch may be enhanced if traps are coloured yellow.
- It has been found that MAT/ BAT closer to the field edges and hedges may be more effective than those in the centre of gourd fields, hence, they should be placed to the edges of field, invariably.

Fig. 20. Hanging of bottle trap

- If ethanol is not available, alteratively, 1 ml methyl eugenol is dripped onto ply wood block of above size evenly and a few drops of malathion may be added. The block prepared, thus, can be used as mentioned above.

As another alternative method, 1.0ml methyl eugenol can be put in 1 liter of water, mixed with 2ml malathion and placed in the orchard/ field in wide mouth bottle at 10 spots. These two alternative methods, however, are short lived and need to be replaced weekly. In certain part of the country, 1.0ml methyl eugenol and 0.1ml malathion on cotton wool is also used as an alternative.

Chemical Control

The spray of insecticides is done to destroy the adults before their maturity and start oviposition. Cover spray kills fly adults on contact and some times destroy eggs and larvae through penetrative action. The residue of insecticide may also keep the flies away for certain days especially when spray is aimed to cover the fruit with insecticide. Many synthetic chemicals, neem and other botanicals have been evaluated and are in use against the fruit flies. Cover spray includes mixture of water and one of the conventional insecticides. In fruit crops, spray can be done once or twice starting from just before the maturity of fruits or if sustained numbers of fly are caught in traps. The amount of insecticide used depends on the size of the tree: larger trees require more insecticide, smaller trees require less. Two insecticidal sprays (30 days prior to harvest - Carbaryl 0.2% and 15 days prior to harvest - Decamethrin 0.0028%) may be done.

Cultural

Field sanitation : Fruit fly lays eggs in fruits and vegetable and the larvae hatch inside the fruit .Generally such fruits fall down to ground and larvae crawl out of the fruit and go into the soil.The hundreds of such larvae can never get a chance to survive if the infested fruits are destroyed. Sanitation is the process of disposing of infested fruits so that the fruit fly larvae do not survive. Sanitation also ensures other methods to be effective. All such infested fruits whether on the tree or fallen on the ground should be collected. Cracked fruits on the tree attract female flies to lay the eggs therefore, damaged fruits, though not infested by fly, should also be removed. Putting fruit residues into compost heaps or a rubbish dump is not recommended. Collected fruits can be destroyed by soaking it in water, topped with a layer of kerosene for 4 -5 days. Fruit fly larvae can survive in water for a few days unless their oxygen supply is cut off. The kerosene layer prevents oxygen exchange between the air and water. Freezing fruit for a few days or cooking fruit (which can then be fed to poultry or pigs) are other methods of disposal. Burial is not recommended since fly larvae can survive burial. If at all, fly-infested fruits are to be buried, first treat them with insecticide. Owners of big orchards realize the benefit of such practice but poor farmers find it difficult to part with such damaged fruits. They sell such fruits in market even if they get the low price. This practice creates problem not only to them but also to nearby orchards. Such farmers need to be educated to follow the practice.

Raking and ploughing : Fruit flies not only pupate inside the soil but also over winter in pupal stage when conditions are not favourable. During winter months the soil in the orchards should be turned over or be ploughed so that pupae underneath are exposed to sunlight and killed. Besides, they also become prey to predators (ants, birds, lizards etc). A large number of pupae are also killed due to mechanical injury during ploughing. Raking can also be done by hand or implement around and below the canopy to a depth of 6 cm .Raking may be done twice in fruit crops- two weeks after colour break and again three week later.

Early harvesting of premature/mature fruits :Avoidance of fruit fly infestation is possible by harvesting crops at a stage of maturity when fruits are not susceptible to fruit fly attack. Bananas, for example, have been exported around the world because they are not susceptible to fruit flies at the mature green stage, except in countries where banana fruit fly and papaya fruit fly occur. In certain parts of country, the fruits are harvested well before the fly population starts damage. However, due to early harvesting growers may not get appropriate market price and fruits may be of inferior quality.

Other Methods

Fruits can be easily protected against fruit flies by bagging them in newspaper bags. The bag provides a physical protection to the fruit by preventing adult female flies laying eggs. Each and every fruit is covered by paper or cloth bag. This is useful when fruit are less in number especially in kitchen garden. Bagging is inexpensive and easy to apply and guarantees nearly complete protection from fruit flies. Fine wire netting of small orchard is another way of protecting fruits from fruit fly. Destruction of non economical and un cultivated wild host of fruit flies is also recommended.

Integrated Control

Integrated management of fruit fly includes trapping (MAT); need based cover spray of insecticide, application of BAT, and cultural control. However the success of these methods depends on grower status (rich *vs* poor), access to technology, cost, scale (single orchard, regional level, national level), globalization of markets, and impact on the environment

- Collection of fallen fruits weekly starting when fruit start maturing. Fruit removed from the orchard should be treated and buried or destroyed and not left exposed. This is labour intensive work but should be done with out fail.

- Soil raking around and below trees to a depth of 6 cm. This should be carried out twice - two weeks after fruit start maturing and three weeks later. Ploughing may be done in winter.
- Use Male Annihilation Technique @ blocks 10/ha using Methyl eugenol. Install the traps at least 45 days prior to fruit maturity. If blocks are wetted in rains or fail to attract the flies replace the blocks after 30 days or even early.
- Use of Bait Application Technique (0.1% insecticide,10% Jaggery or banana in water) especially when fly pressure is high . Apply BAT mixture weekly starting 45 days prior to fruit maturity. Spray in spots of 40 ml at a rate of 200 spots/ha (approximately 8 L/ha). Apply bait spray in spots to surroundings hedges also.
- The above mentioned schedule will ensure the fruit fly control in any area, however, if area is highly fruit fly prone, insecticidal sprays (30 days prior to harvest - Carbaryl 0.2 or Decamethrin 0.0028%) at 15 days prior to harvest can be applied.
- Avoid delay in harvesting. If needed, do post-harvest hot water treatment within 24 hr. after harvest.
- The fruit flies attack different crops in the same area, their management must be based not only on mango, but also on wild hosts and commercially grown host plants. Thus, to improve the efficiency of fruit fly management, fruit growers should pursue a concerted fruit fly population management strategy over a significantly large area.

Post-Harvest Disinfestations

There are two safer options for post-harvest disinfestations of mango fruits against Fruit fly. (a)After proper harvesting select uniform sized undamaged fruits. Dip them in 5% solution of sodium chloride in cold water for one hour. This will kill 95% eggs in fruit epicarp and also remove the externally present pesticide residues. However, this can't meet international quarantine standard (99.999% freedom).(b)Post-harvest immersion of mango fruits in hot water at 48 0C for 45-60 minutes give 100% mortality of fruit fly eggs in the epicarp without affecting fruit quality. The hot water treatment should be done at very small scale initially to develop expertise. A little lapse may cause hardening of fruit.

Fruit Borer (*Deonalis albizonalis*)

It is a major pest in Orissa and coastal Andhra Pradesh. The pest is active from January to May. Adults lay eggs on fruits and hatched young caterpillars attack fruit from marble size onwards, bore into the fruit at the bottom (beak region) and feed inside reaching kernels. They can make series of tunnel in kernel. Infested fruit may rot and fall prematurely. A single larva can damage many small fruits. Maturing fruits may also be attacked. The characteristic symptom of damage is appearance of small hole with dark pale brown ring encircling it. Entrance hole is plugged with excreta which may be used for pupation (mostly pupation may takes place out side) and exit of emerged moth. Fully grown caterpillars (25 mm) have red bands on body alternating with white bands. There are usually 3-4 generations in a crop season. Larvae of the last generation hibernate in the dead wood. Castor shoot and capsule borer, is also seen boring into fruits, when two fruits are seen attached together.

Management

- Collection and destruction of fallen fruits and dried branches/ dead wood after fruit harvest and during off season even though laborious, reduces infestation in the following season.
- Collection and destruction of infested fruits regularly stating from the beginning of the appearance of the pest to prevent the spread of pest infestation.
- Collection and destruction of diapausing prepupae from the crakes and crevices of the bark during off season.
- The mango fruit borer larvae/pupae/diapausing larvae can be trapped by using artificial niches on the tree during the season and destroyed.
- Swabbing of the chlorpyriphos 20% EC @5 ml/lit. of water on tree trunk would kill the prepupae/ pupae population under the bark and helps in reduction of fruit damage.
- Spraying of insecticides such as chlorpyriphos 20 % EC (2.5ml/l of water or dichlorvos 76%EC (1.5ml/l) in the second fortnight of January coinciding with the moth emergence/hatching of eggs of first brood in the gardens where the pest incidence was severe in previous year. Spray Endosufan 0.07% at marble size onwards and repeat with Deltamethrin 0.0028% after two weeks in case of heavy infestation, or

- Application of dichlorvos 76%EC (1.5 ml/l) or chlorpyriphos 20%(2.5ml/l)or carbaryl 50% WP (3g/l)or neem oil 3ml+Chlorpyriphos 1ml/l late in the evening when fruits are of marble size to coincide with the migration of the larvae.

Mango Stone Weevil (*Sternochetus mangiferae*)

Sternochetus mangiferae causes direct damage to mango fruits varying from 5- 80 percent depending on the place and variety of mango. The loss is essentially in four folds: One-Affected fruits prematurely fall. Two- Infested fruit results in spoilage of pulp, Three-The weevil has been hindrance in export of mango fruits and Four- weevil damaged seeds may limit plant propagation in nurseries and, orchards. Three species of stone weevil, namely, *Sternochetus mangiferae, S. gravis* and *S. frigidus*have been reported as pests of mango. This pest is supposed to be monophagus. Weevils are more in humid areas, and near coastal belts, however, the weevil is found even in the South Indian plains and moderate areas.

Adults become reproductively active when mango plants begin to flower in about mid March and reach a peak during the first week of April. The oviposition period varies from 3 to 5 weeks. Females select fruit at random and lay eggs singly on the skin or sometimes on the stems; most eggs are laid on the sinus of the fruit. The female makes a boat shaped cavity in the epicarp into which an egg is deposited. Incisions made by ovipositing females are small and generally soon heal.

The grubs after hatching from eggs, enter the nut or stone. Initially damage is caused by feeding on the outer coat of the stone in a zig-zag fashion. Complete larval development usually occurs within the maturing seed, but also very occasionally within the flesh. Generally, only a single larva completes development in each fruit, but as many as five larvae have been found. Consequently, complete stone is destroyed leaving behind black frass. Damaged stones lose their viability. Fruits are also rendered unfit for processing or direct consumption due to mixing of excreta by the weevils while emerging from ripe fruits. No external symptoms are readily visible on infested fruits. As fruit and seed develop, the tunnel and seed entry are completely obliterated so that in time it is impossible to distinguish infested and non-infested seeds unless they are cut open. Yields are not significantly affected, since the larvae usually feed entirely within the stone.However, post harvest damage to the pulp of late-maturing cultivars by emerging adults occurs. Pupation takes place with in the kernel. Seeds, in which the embryo is damaged and the reserve of food in the cotyledons is greatly reduced, fail to germinate. Probably its greatest significance as a pest is to reduce the germination capacity of seeds greatly and to interfere with the export of fruit, because of quarantine restrictions imposed by importing countries. In India, all cultivars are susceptible and levels of infestation vary.

The mango weevil produces only a single generation per annum. They have well developed wings and fly readily. After emergence, adults enter a diapause, the duration of which varies with the geographic range.During non-fruiting periods, weevils diapause under loose bark on mango tree trunks and in branch terminals or in crevices near mango trees. The weevils over season under bark and stone walls, where they remain dormant until the next flowering season. Adults usually remain in the vicinity of the parent tree until the following fruiting season and high infestations appear year after year in some locations and low infestations appear in others. Longer range dispersal occurs largely through transport of fruit and seeds containing the larvae, pupae or adults.

Management

- In nursery beds, far more seeds may be sown than are required for the projected number of seedlings to overcome a low percentage of germination. Alternatively, the seed may be shelled and only sound kernels planted.

- The mango weevil has few natural enemies. Parasitoids are unknown, probably because of the concealed nature of most of the weevil's life stages. Adults may be susceptible to predation by ants, rodents, lizards and birds.

- The removal of all fallen fruits and seeds is very labor-intensive and demands complete removal and disposal of fallen fruits from affected orchards. However, studies have shown that fallen fruits between April and May have grubs, which die after fruit fall. Fruits that fall a month prior to harvest may harbour pupae, which are crucial, as adults who emerge can carry over to next season. These especially should be removed. This procedure has been only partially effective for pest control. During harvest, a small proportion of fruit show ripening on the tree.These are invariably eaten by labourers and stones are thrown in the orchard itself. This should be discouraged, as the seed may contain weevils, which will emerge from the seed and remain in the orchard till next season.

- Raking of soil below the tree can effectively contribute partially to weevil management. Keep the tree basins clean to prevent hiding of adult weevils.
- Sticky bands should be applied at upper end of tree trunk to prevent migration of weevils to branches for egg laying on fruits during February.
- Spot application of fenthion (0.05%) or carbaryl (0.1%) during off season (December - January) on tree trunks up to the height of two metres should be done.
- If failed to do spot application, spray deltamethrin (0.0025%) six weeks after fruit set and 15 days thereafter, spray fenthion (0.05%).

Fruit Sucking Moths (*Othries* sp., *Achaea* sp.)

Fruit-piercing moth attacks many fruit and vegetable crops including mango. Adult moth is the damaging stage, and the larvae are essentially not harmful. The mouth parts of the moth are about an inch (2.5 cm) long and strong enough to penetrate through tough skinned fruit. Once the moth has punctured the skin of the fruit, a process that usually takes a few seconds, it feeds upon the juices of the fruit. Feeding occurs at night and the fruit does not have to be ripe to be fed upon by this moth . Fruit flesh damaged by this moth becomes soft and mushy. Damage caused by this pest is not only a result of the direct feeding of this moth but also by the fungal and bacterial infections that develop at the wound site. When moths are abundant green fruit is attacked, causing premature ripening and dropping of fruits. Incidence of damage by this moth is normally low, however when outbreaks occur, most of the crop is affected.

Eggs are generally laid on the underside of leaves of weeds and bushes but may be found on the mango, bark or on other plants nearby. Eggs hatch in 3 to 4 days. Caterpillars are cylindrical in shape and are 1/5 to 1/3 inch long during the first larval stage. They mostly feed beneath or on the edges of leaves. Young larvae drop to the ground at any sign of danger, while the older larvae take an aggressive attitude by hanging on to the food plant with their hind legs and swaying the rest of their body from side to side. They reach up to two inches in length when fully grown. They are either dark green to black or pale green to yellow.

Mature larvae pupate within a cocoon spun between leaves and woven together with silk. The leaves containing the cocoon may remain on the host plant or dry and fall to the ground. The adult moth is large and robust. It has a wing span of almost 10 cm and a stout body, about 5 cm long that does not extend, or slightly extends, beyond the hind wings. The eyes are large. The area behind the head of the moth, the thorax, is pale to purple-brown and the abdomen is pale brown at the base brightening to yellow-orange at the tip. The forewings resemble a leaf by being olive to purple-brown and may have white and green colored flecks (the colored flecks are more common on females). This leaf-like appearance of the forewings makes this moth difficult to see when it is at rest, especially, because the bright hind wings are not visible. Although their flight is slow and heavy, the adult moths are very strong fliers and can travel great distances from their breeding grounds in search of food. Adults fly mainly between the hours of dusk and 11:00 PM. They are not usually attracted to light.

Management

Capture and destruction of Moths : Since this moth is not easily disturbed from fruit once it has begun to feed, netting and killing of moths is possible. This is best accomplished an hour after sunset when there is sufficient darkness with the aid of torches or a strong flashlight. This method is most feasible when fruit are easily accessible and populations of this moth are small. This method is not very effective once a large population of this moth exists and is usually recommended only if no other control means is possible.

Illumination of orchard : This moth prefers darkness and avoids light, therefore, pest causes less severe damage under illuminated conditions, however, this is an impractical means of reducing damage.

Bagging or screening fruits : Mechanical protection of fruit against moth attack is achieved by covering the fruit with brown paper or transparent oil paper bags. Brown paper bags last for about a month in the field. This method is not much practical in mango as fruits are small, clustered or scattered, and not within easy reach.

Smoking of the orchard : This method involves the obscuring of the odor of mature and ripening fruit that attracts the moth with the odor of the smoke. Containers full of inflammable material, oil, tar and some green plant trimmings to enhance the smoke is placed within the orchard at a rate of 2 to 4 per acre. The smoking process is started a half an hour before dusk and continued for 2 to 3 hours after nightfall. This period represents the time in which the moths are seeking their night time feeding grounds. If the smell of the orchard is masked, the moths choose various wild hosts and remain on them throughout the night to feed.

Orchard sanitation : This method involves the regular collection and proper disposal of all attacked and spoiled fruit. Both fallen fruit and attacked fruit on the tree should be collected, buried deeply for composting. These procedures dissipate the odor emanating from the spoiled fruit so they cannot serve as an attractant for the moths.

Forced harvesting : When severe infestations are anticipated, damage may be avoided by harvesting the entire crop as soon as signs of fruit ripening is observed. The grower must determine if the revenue loss from premature harvesting is less than that from possible damage.

Chemical Control : The effectiveness of chemical control is variable and remains ineffective in most of the cases.

Pests of Mango Flower

Midges, caterpillars, leafhoppers, thrips and mites are the important pests attacking mango inflorescences.

Mango Hoppers (*Idioscopus sp.* and *Amritodus atkinsoni*)

Among all the insect pests that attack mango, hopper is considered as the most serious and widespread pest throughout the country. Some 18 species of leafhoppers have been reported as pests of mango in the world, of which fifteen species have been reported on mango from Asia. Of these, *Idioscopus clypealis* (Leth.), *I. niveosparsus* (Leth.), *I. nagpurensis* Pruthi and *Amritodus atkinsoni* (Leth.) are important species.

The mango leaf hoppers are small wedge shaped insects and measure 3.4 - 5.1 mm in length. *Amritodus atkinsoni* is comparatively bigger and darker while *I. clypealis* is smallest and lighter in colour. A low population of hoppers has been recorded in mango orchards throughout the year but it shoots up during February-April and June-August. It normally oviposits in mango inflorescences during Feb.-March. The female hopper lays 100-200 eggs on mid rib of tender leaves, buds and inflorescence. Development from egg to adult normally lasts 25–30 days. Nymphs feed on inflorescences during March and April. From May to February of the following year, only aestivating adults are found.

Due to feeding injury of large number of nymph and adult hoppers, infested flowers shrivel, turn brown, become dehydrated, and fruit set does not occur as they ultimately dry and fall off. The extent of damage depends upon the critical crop stage and hopper population. Leaf hoppers' population on inflorescence has significant negative correlation with fruit set.They also damage the crop by secreting a sweet sticky substance which encourages the development of the fungus *Maliola mangiferae,* commonly known as sooty mould which adversely affects the photosynthetic activities of the leaves. Leafhoppers love damp and shady places during summer and their populations are quite high in the old, neglected and dense orchards. Populations fluctuate throughout the year, but peaks coincide with vegetative and reproductive phases of mango trees. Temperature and relative humidity constitute important environmental factors regulating the population of the leaf hoppers.

Management

- In old dense orchards, where tree canopy has become dense, prune some branches to have better light penetration. Avoid high density planting with present commercial varieties. Several intermittent flushes are noticed due to regular irrigation and fertilization in some regions. These flushes support several generations of leaf hoppers in a year. Regulate these flushes by reducing the inputs and pruning and establish the orchard with a proper canopy management plan.
- A number of natural enemies of leafhoppers occur among them two entomopathogens, *Verticillium lecanii* and *Beauveria bassiana* are known to infect *I. clypealis*. These are available in the market and can be used.
- Coccinellid beetles are most common predators on nymphs of leaf hopper. Besides these, spiders and reduviid also keep hopper population in check .Conserve them by avoiding sprays of broad spectrum pesticides during their peak activity period.
- Spraying of 0.2 per cent carbaryl in mango orchard during off season (November) especially on trunks, helps in delaying the

development of nymphal population during flowering period. This may help in avoiding the insecticidal spraying during flowering and fruit setting, and thereby protecting the pollinators in mango orchard which are active during this period.

- Fogging orchard with a 1.5:8.5 of malathion and Diesel mixture has been found effective, cheaper and less time consuming than conventional spraying or
- Successful control of hoppers can be achieved by integrating chemical insecticides with neem products depending on hopper density. First spray should be given at early panicle emergence with the following:

(i) At high hopper density (>4 hoppers/panicle) spray Imidachloprid 0.005% or Lamda cyhalothrin 0.0025%

(ii) At low hopper density (<4 hopper/panicle) spray Azadirachtin 3000 ppm 2 ml/ l or 10,000 ppm 1 ml/l .

The second spray may be done with 0.15% Carbaryl at full length stage of panicles but before full bloom to avoid killing of pollinators. The third spray if required should be done at pea size fruit with Imidachloprid or Lamda cyhalothrin (as above).

Mango Gall Midges

Fourteen types of galls; eleven on the leaf, one in the leaf axil, one on the inflorescence and one on the stem have been found to occur in mango. Of these *Erosomya indica, Dasineura amaramanjarae, Procystiphora mangiferae, Amradiplosis allahabadensis, Procontarinia mangifoliae and Procontarinia matteiana* are important.

The mango gall midge or mango blister midge *Erosomya mangiferae* Felt, is a major pest, destroying flowers and up to 70% of set fruit. The pest is commonly known as mango blossom midge, mango gall midge, mango inflorescence midge, mango blister midge, mango blister and mango midge.

The adult midge is harmless minute fly which is short lived and dies within 24 hours of emergence after copulation and oviposition. The flies lay eggs singly on floral parts like tender inflorescence axis, newly set fruit or tender leaves encircling the inflorescence. The eggs hatch within 2-3 days. The midge infests and damages the crop in three different stages.

i) The first attack is at the floral bud burst stage. Eggs are laid in newly emerging inflorescence, between folds of sepals and petals of the flower buds. Eggs hatch and larvae tunnel the axis which prevents flower opening and thus destroy the inflorescence completely. Infested panicles have a characteristic right-angled bend, with a soiled exit hole, from which last instar maggots emerge. It finally dries up before flowering and fruit setting. The mature larvae make small exit holes in the axis of the inflorescence and slip down into the soil for pupation. The inflorescence shows stunted growth and its axis bends at the entrance point of the larvae.

ii) The second attack of the midge takes place at fruit set. The eggs are laid on the newly set fruits and the young maggots bore into these tender fruits, which slowly turn yellow which eventually drop before the marble stage.

iii) The third attack is on tender new leaves encircling the inflorescence. The most damaging one is the first attack in which the entire inflorescence is destroyed even before flowering and fruiting.

The larval and pupal period varies from 7-10 and from 5-7 days, respectively. There are 3-4 overlapping generations of the pest spread over the period from January-March. Thereafter, as the weather conditions turn unfavourable, the mature larvae undergo diapause in the soil instead of pupating. They break diapause on the arrival of favourable conditions in following January. Climatic factors such as rainfall, sunshine, temperature and humidity affect gall midge populations in the field. Gall midge adults may disperse by wind or larvae may be transported with soil or other substrates covering the roots of mango plants in nurseries.

Management

- As the larvae pupate in the soil, ploughing of the orchards expose pupating as well as diapausing larvae to sun's heat which kills them.

- Soil application of carbaryl dust also kills pupating as well as diapausing larvae in the soil. The insecticide in the soil should be applied after monitoring larval population on white sheet below the tree.

- Spraying of 0.05 per cent Fenetrothion or 0.045 per cent Dimethoate or 0.04 per cent Diazinon at the bud burst stage of the inflorescence has been found effective in controlling the pest population.

Pests of Mango Buds and Leaves

Foliage feeders are one of the largest groups of injurious insects of mango. Pests of mango buds and foliage may cause damage by reducing the photosynthetic area of the plant, thereby the quantity of photosynthates translocated to the root and back to the fruit is reduced. The most destructive mango leaf feeders are thrips, midges, mites, scales, whiteflies, mealy bugs, weevils, ants, and caterpillars.

Leaf Cutting Weevil (*Deporaus marginatus*)

It is about 5 mm long and 2 mm wide. Before egg laying the female surveys the whole leaf surface and then moves to dorsal side and makes a minute cut by the side of midrib and prepares a 'c' shaped pouch for an egg. The female then turns and bend the abdomen to lay an egg inside the pouch. Weevil cuts the leaf on tree leaving about 1/4 of it on the tree and the cut appears as if it is cut by a sharp blade. The fallen leaf pieces can be seen under the tree. The weevil feed by eating away small patches of epidermis, the injured area turns brown and the leaf curls up or crumples. Seedling as well as young trees suffer more than the older ones. Due to this damage, growth of plants may be affected. It is very active during rainy season and cause severe damage in nursery plants.

Management

- Remove and destroy all fallen pieces of leaves.
- Cover the nursery plants with nylon net.
- Hoeing and ploughing below the plants help in population reduction of this pest.
- Drenching the plant basin with 0.2% carbaryl dust also reduces the pest pupulation.
- Spray of dichlovas 0.05% or endosulfan 0.1% or dimethoate 0.03%.

Mango Mealybug

More than 20 species of mealybugs are known to attack mango. Of these, *Drosicha mangiferae* is a serious pests.

Fig. 21. Polythene Band

Drosicha mangiferae Green is the most common mealy bug and causes severe damage to mango crop throughout the country. The late instar nymphs and adult female bugs are flat, oval and waxy white. The adult male is winged and small, female is bigger and wingless. The female, after copulation, crawl down the tree in the month of April-May and enter in the cracks in the soil for laying eggs in large numbers encased in white egg sacs. The eggs lie in diapause state in the soil till the return of the favourable conditions in the month of November - December. Just after hatching, the minute pink to brown coloured nymphs crawl up the tree. The population peak is observed during March-April on inflorescence. These migration periods are critical from management point of view. After climbing up the tree, nymphs and female bugs suck sap from inflorescence, tender leaves, shoots and fruit peduncles. Excessive and continuous draining of plant sap causes wilting and finally drying of infested tissue. They also secrete honey dew, a sticky substance, which encourages the development of a fungus *Maliola mangiferae,* termed as sooty mould.

These are polyphagous and infest more than 70 plant species. They are considered more important because they infest the crop during the flowering season and if the control measures are not taken timely , the crop may be destroyed completely. It is univoltine. The intensity of attack of this mealy bug varies from year to year and locality to locality in India, probably because of soil and environmental conditions. Moderate rainfall (55–60 mm) during oviposition and dry conditions during hatching appear to be favorable for development.

Management

- The IPM schedule of mealy bug is very important and useful if timely operations are done. 11 species of coccinellids, six species of spiders, two species of mites and three species of parasitoids have been found attacking this pest. Conservation of these natural enemies may be done by avoiding use of broad spectrum insecticide during their peak activity period and minimum tillage.

- Among the pathogens, the entomogenous fungus, *Beuveria bassiana* (10×10^7spores/ml) gives complete mortality of nymphs within 5-12 days.

- Flooding of orchards with water in the month of October kills the egg and all the branches that touch ground should be removed. Weeds like *Clerodendrum infortunatum* which is additional host of this bug may be removed. Further, ploughing the orchards may be done in the month of November to expose the eggs to sun heat.
- In the middle of December, 400 gauge alkathene sheet of 25 cm width may be fastened to the tree trunk besides raking the soil around the tree trunk and mixing of carbaryl dust. The dust may also be sprinkled below the atkathene band on the tree. The congregated nymphs below the band may be killed by any of the suggested insecticides.
- In case the nymphs have succeeded in crawling spraying of 0.2 per cent Carbaryl have been found useful in controlling early instar nymphs.

Leaf Eating Caterpillars

There are number of caterpillars that attack mango leaves and destroy them. Some of them are discussed below

Slug caterpillar (*Latoia lepida)* : It is most common in the southern regions of the country. This pest damages the castor plant sporadically. The female moth lays flat scaly eggs on the tender parts of the plant in small clusters. The full-grown caterpillar measures 15-25 mm in length. Thick short spiny hairs out of rows are found on the body. It is bright green in colour and has interrupted blue stripes on the dorsum. It moves like a slug. Pupation takes place in a hard greyish cocoon on stem or the trunks of its other host plants. The pupal period lasts from three to five weeks or longer. The caterpillars are spiny and when touched, cause irritation to the skin. To begin with, they feed gregariously on the leaves and later spread over to the entire plant.

Orgyia postica : The female moth lays eggs in groups. The full grown larva is a light brown hairy caterpillar with orange red mouth and creamish yellow and brown tufts of hair. The young larvae feed on new flush and nibble the shoot tips. The caterpillars can cause large scale defoliation of mango trees and also feed on the stalk, skin and pulp of fruits. Larvae cause more severe damage to fruit than leaves, causing the fruit to drop prematurely or show obvious signs of damage on the skin and pulp. Pupation takes place on the plant in a thin , papery cocoon covered with larval hairs.

Euthalia garuda garuda: It is a medium-sized dark brown beautiful butterfly which flies with stiff wing beats and often glides. The caterpillars are bright green about 3 mm in length with stout long hairs arising from each segments of the body. The pupa is green and quadrangular and found glued to the leaf from distal end. The caterpillars feed on leaves and defoliate the plant and make large number of plants leafless. Severe defoliation affects vigour and vitality of tree ultimately flowering and fruiting.

Lymantria marginata: Moths are small to medium size with hairy body. They are short lived, mouth parts are poorly developed and do not feed. They hold their board wings like roots over their abdomen at rest. Some species adults rest with wings flattened forming a triangle and the fore legs extended forward. They are usually active at night. The eggs are laid without any cover of hair in clusters on the surface of the bark crevices. Eggs are oval and grey in colour. The newly hatched caterpillars congregate near the egg mass and feed on empty egg shell and later on leaves. Brown color body of caterpillars changes to grayish black later on. Caterpillars are usually hairy, ash colour, often with four distinct tussocks of hair on their back make them look like a tooth brush. The hair can cause irritation on contact. The caterpillars pupate in a woven cocoon under bark or foliage. The pupa usually covered with larval hairs as well. These hairs can cause problems if touched, or if they are released to blow about.

Management

Many of the caterpillar cause irritation in skin hence, they should be picked carefully in the gregarious stage and their cocoon should be collected which are found in numbers on tree trunks. These may be destroyed in water with a film of kerosene.

Spray Carbaryl 0.2% or Endosulphan 0.07%, if pest attack is more.

Shoot Borer (*Chlumetia transversa*)

The shoot borer has been observed causing serious damage to new flush and new panicles of mango throughout the country. This pest is usually active from August to October. The adult moths are shining grey in colour and measure about 17.5 mm with expanded wings. Hind wings are light in colour. Female moths lay eggs on tender leaves. Eggs are oval in shape and pale yellow in colour. After hatching, young larvae enter the midrib of leaves and then enter into young shoots. Full grown larva is dark pink with dirty spots and measures 24-24 mm in length. New shoots are damaged by the caterpillars by tunnelling from growing tip downward. As a result, the shoots droop and finally dry and wither. Larvae also bore into the inflorescence stalk. When full grown , these caterpillars come out and enter into the slits and cracks in the bark of the tree, dried malformed inflorescence or cracks or crevices in the soil for pupation. Upper and lower parts of the plants are more infested than middle parts. The affected plants become stunted. Young trees up to the age of 8-10 years are more susceptible. There are four overlapping generations of the pest in a year and it overwinters in pupal stage.

Management

- The attacked shoots may be clipped off and destroyed.
- Spraying of Carbaryl 0.2% or Quinalphos 0.05% or fenitrothion 0.047% or 0.05% malathion at fortnightly intervals from the commencement of new flush gives effective control of the pest. A total of 2-3 sprays may be done depending on the intensity of infestation.

Leaf Miner *(Acrocercops syngramm)*

This pest has been reported from Bihar, Bengal, Maharashtra, Karnataka, Andhra Pradesh and Tamil Nadu as a minor pest. This pest is usually active from October/ November to April/May. Adults are very small, delicate months with narrow long fringed wings. Caterpillars are legless and pale greenish-yellow. Tiny caterpillars mine under the dorsal epidermis of tender leaves and feed within, as a result greyish-white blisters appear on leaves. Caterpillars pupate with in the tunnels.

Management

- Remove and destroy promptly the affected leaves in the initial stage of attack. This will prevent the pest infestation from spreading.
- Spray Cypermethrin 0.0125% or endosulfan 0.07% during new flush emergence.

Aphids

Aphids have a very wide host range. Most severe attack by aphids is usually observed in young plantations; young shoots are densely colonized in spring. Leaves colonized by aphids are slightly rolled at the succulent shoot apices and usually curve downwards at the tip. The black aphid is more rarely found on flowers and flower buds. In severe infestation, the entire new growth on young trees can be destroyed i.e. shoot tip.

Management

- There are many predators of aphids that feed on aphids and reduce the population. Try to conserve them.
- Application of dimethoate 0.03% can control the aphids. Early applications, when the aphids are fully exposed and before they curl the leaves, are most effective.

Flea Beetle (*Rhynchaenus mangiferae*)

Mango flea weevil or mango leaf mining weevil infesting mango crop is widely distributed and is a serious pest of mango at the time of flower emergence. The population of weevil is found more in on year than off year indicating that it infest mainly the flowers. Adult activity begins in the February and continues until the end of April. The adult population peaks in the March on panicles and in the April on new leaves. Egg laying takes place during the day, females lay eggs in slits made in the florets. Grubs eat the entire flower except sepals which keep the larva, compact and covered. Only one larva or pupa is found in a flower bud and the adult later comes out through the exit hole. The adult feed on panicle, flowers and fruits and some times cause severe damage. The adult population goes to the new flush of leaves after fruit set. The egg period lasts 2 days, while the larval and pupal periods both last for 4 days. Adults are quiescent from April to February.

Management

- The affected fallen leaves may be collected and destroyed. The orchard may be kept clean, weeds may be removed regularly. Hoeing or ploughing the soil may reduce the infestation.
- Fenvelrate 0.005%, cypermethrin 0.005%, endosulfan 0.07% are effective insecticides against this pest.The spray should be done

early in the flowering season. The spray done for hoppers may control this pest also.

Grey Weevil (*Myllocerus* species)

They are polyphagus pests. Adults are poor fliers but are very active feeders on the leaves of mango. *M.discolor* has long snout with grey color. It attacks leaves, shoot and flower. In winter they are not active. Adult weevils congregate on the tender leaves and nibble irregular holes on the leaves and some times consumes the entire leaf leaving the midrib only. Grubs feed on roots.

Management

Hand picking of the adult weevils and spraying of 0.07% endosulfan will solve the problem.

Mango Shoot Gall Psylla (*Apsylla cistellata*)

It is a very serious pest of mango in plains of northern India (Terai region of U.P., north Bihar and West Bengal), Nepal and Bangladesh. The profuse egg laying (nearly150 eggs) is found in the midribs as well as in lateral axis of new leaves. The incubation period has been reported 191-211 days. Nymphs emerge from eggs during August-September and crawl to the adjacent buds to suck cell sap . As a result of feeding, the buds develop into hard conical green galls. The galls are usually seen during September-October. Consequently, there is no fruit set. Nymphs pass the winter inside the galls. Adult psyllids are with black head and thorax and light brown abdomen and measure 1.72 mm (male) and 2.50 mm (female). The activity of the pest starts from August. The adults emerge from the galls from the third week of February till third week of March .The galls dry out after emergence of psyllid adults in March. The insect has only one generation in a year.

Management

- Pruning of infested shoots and their burning in the month of April to July is effective. The galls with nymphs inside should be collected and destroyed to prevent carry over of the pest.
- Raking of the soil around the base of the trees in January, after the last generation has pupated, helps in checking the pest population.

- The pest can effectively be controlled by spraying Dimethoate 0.06% or Quinalphos 0.05 % at 2 week intervals starting from the middle of August. The use of same chemical for every spray should be avoided.

- Removal of egg bearing leaves during June and pruning of gall bearing shoots during January is also recommended.

Thrips

(Rhipiphorothrips cruentatus, Scirtothrips dorsalis, Haplotrips sp, *Thrips palmi* and *S. mangiferae)*

Thrips are widely distributed and polyphagous. Generally thrips are of minor importance but occasionally serious. Adults are minute, slender, soft bodied insects with heavily fringed wings. They attack leaves, inflorescence and fruits. Nymphs and adults lacerate the tissues and suck the oozing cell sap. Leaf feeding species would emerge when tender leave are ready as ovipositional sites. The nymphs and adults rasp the ventral surface of the tender leaves and flower stalks and suck the oozing sap. Affected leaves show silvery sheen and bear small spots of faecal matter. In case of severe infestation young leaves remain small, leaf tips turn brown and get curled. The twigs of infested shoots are much shorter than those of un infested ones. Inflorescence and young fruits when infested show discoloured tissues at the feeding site, which subsequently turn brown. The population of the thrips is low during winter, increases in early spring and reaches its peak during summer .

Management

- Yellow sticky traps can be used for monitoring thrips densities.
- Remove all the pruned materials from field and burn and rake the soil periodically.
- Spray or Acephate 0.075% followed by Dimethoate 0.06%+ Azadirachtin 2ml /1 where thrips are major problem.

Leaf Gall Midge (*Procontarinia matteriana*)

Adult lays eggs on the underside of the leaves during March, July and October. On hatching, maggots bore inside leaf tissue, and feed within, resulting in formation of small raised wart-like galls on leaves. Affected leaves get deformed and drop prematurely. The other species cause minute pustular galls or round, echinate galls on the leaves. The same leaf may have different types of galls. The number of galls varies greatly from a few to 700 and in cases of heavy attack the leaves are entirely covered with bead-like structures. The leaves are rendered useless by continuous draining of the sap by larvae feeding inside the gall.

Management

- No definite control measures against mango leaf gall makers have been evolved as yet. Some of their natural enemies have been recorded. If infestation is severe, especially in young orchards, spray of dimethoate 0.03%, may be done.

Red Tree Ant (*Oecophylla smaragdina*)

Red tree ant are found throughout India. The ants web and stitch together a few leaves, usually at the top of the branches and build their nests. The ants are carnivorous and prey upon small insects. However, indirect damage is caused by protecting insects like aphids and scales, which excrete honeydew. Besides, being ferocious, may also be of nuisance to workers around.

Management

- Nests should be removed and destroyed mechanically or by spraying any of the contact insecticides, after disturbing the nest.

Mango Scale Insects / Coccids

Coccids are found throughout India. They attack many tropical and subtropical fruits. The oval, immobile, brown scale insects are frequently grouped near the tips of young shoots, on fruit stalks and on the fruit. Hard scale insects or armored scales are small, about 3 mm long. Their colors vary from white, yellow, gray, reddish or purplish. Adults are oval or round, hard bumps, and sometimes bearing a nipple or dimple in the center. They do not produce honeydews but they secrete armor wax in an oyster shell or circular pattern. They remain attached to the host plant when shells are lifted up. Soft scales eggs are covered or found under the mother scales. Eggs hatch into crawlers (nymphs), which are flattened and looking like dusts on the plant surface. Adult females are either oval or round, soft, legless bumps, and are wingless. Males are tiny yellow-winged soft scales. Soft scales secrete honeydews which attract ants. The shells of the soft scales are not left on the plant when lifted up. The soft covering they secrete, cannot be separated from the scale's body. Soft scales typically move between branches and leaves during their life cycle. Scales remove plant saps by using their sucking mouthparts. Leaves become stunted and turn yellow and the twigs and branches dieback. Some scale species produce honeydews which they secrete while feeding. Honeydews on fruits, leaves, twigs,branches, and barks attract black sooty molds to feed and grow, causing these plant parts to blacken. Adults and nymphs suck the sap of shoots, leaves, fruit stalks and fruit. They weaken plants by puncturing the tissues and consuming sap, but the major damage is caused by the production of large amounts of honeydew upon which saprophytic fungi develop. The resultant thick black layer of sooty mould causes a drastic reduction in photosynthetic efficiency, resulting in premature leaf drop. Fruits heavily infested with scale are unmarketable. They are attended by ants, which probably provides protection from parasites and predators. Several species of scale insects attack the mango in India. Of these five are most serious: *Aspidiotus destructor* Sign. *Pulvinaria polygonata* Cock., *Parlatoria pergandii* Comst., *P.cinerea* Hadden. and *Lepidosaphes gloverii* Pack. *A.destructor* occurs in the mango-growing tracts of Uttar Pradesh and the Punjab; *Parlatoria cinerea* and *Pulvinaria polygonata* in Uttar Pradesh, and *Parlatoria mangiferae* in Orissa. *A.dectyospermi* Morgan, *Diaspis mangiferae* Green, and *Lecanium bicruciatum* Green occur in S. India. *L.gloverii* occurs in Uttar Pradesh only.

Management

- Pruning of infested branches specially in summer and burning them can be resorted to for the control of small scale infestation.
- Around 40 species of predators and parasites have been reported to attack this scales.
- Chemicals used on scales include diazinon, dimethoate, formothion, and malathion. Sprays are only effective on the crawler stage of scales. However, control is difficult on other life stages. Adults are firmly attached to the plant and remain so after their death that may give a false impression of the pest status. Chemical applications should be used only when parasites are not economically effective. The application of pesticides may kill natural enemies of the scale and result in a resurgence of the pest.

Black Fly (*Aleurocanthus woglumi*)

Aleurocanthus woglumi Ashby is found abundantly on various Citrus species, guava, mango, cashew nut, annona and rose. Adults are seldom over 1 mm long, aleurodid type with bluish pubescence on body. They lay eggs on the underside of leaves in a typically spiral pattern. Nymphs are sessile, with black straight bristles. Third nymphal instar is black with whitish waxy secretions on edge of body. Populations are more abundant and more widely distributed during the rainy season. Moderately high temperature and relative humidity are favorable for development. The whole life cycle can be completed in 60-72 days. There is considerable overlapping of generations, so that all stages may be present throughout the year. However, it is during the rainy season when their activity and damage reach peak. It is a minor pest of mango in India. The nymphs and adults suck the sap of leaves and render the trees weak. Nymphs produce copious honeydew, which encourages sooty molds on upper surfaces of leaves, followed by gradual weakening of the tree. The fruit is stained by the 'soot'. Heavy black colonies of the pest occur on the underside of leaves. Its life history on the mango tree has not been worked out.

Management

- Predators can eliminate tremendous amounts of nymphs and pathogenic fungi as *Aschersonia aleyrodis*, are also important biotic mortality factors. Parasites as *Prospaltella* sp. are very effective in controlling the pest.
- Close planting, water logging or any other such condition should be avoided.
- In case of localised infestation, the affected shoots should be clipped -off and destroyed.
- Excessive irrigation,application of nitrogen and pesticidal sprays should be avoided.
- Spray Dimethoate 0.05% during new flush.
- Avoid over spray of insecticides.

Tea Mosquito Bug (*Helopeltis antonii*)

Major pest of cahsew, occasionally damages mango and other fruit crops. Found in Kerala, Karnataka, Goa, Maharashtra, Tamil Nadu and

Andhra Pradesh. Adult is a reddish-brown bug with black head, red thorax, and black and white abdomen. A knobbed process arises mid dorsally on thorax. Eggs are inserted into epidermis of tender shoots and axis of inflorescence. Adults and nymphs feed on petioles, tender shoots and leaf veins causing necrotic lesions. Drying of shoots, inflorescence and dropping of the flowers and young fruits is seen.

Management

- If infestation is severe, spray insecticides like endosulfan 0.07% or carbaryl 0.15%

Mango Leaf Webbers (Orthaga species)

The adult moth is medium sized (31 mm wing span) and having grey body with yellowish white fore wings. Its infestation starts from the month of April and goes up to December. Eggs are laid singly or in clusters within silken webbings on leaves. Eggs are dull greenish in colour and oval-shaped. Upon hatching, the caterpillars feed on leaf surface by scrapping. Mature larvae are pale grey with prominent white stripes on the dorsum and measure about 25 mm in length. Larvae are very active and the moment touched, they fall to the ground. The larvae web the leaves and the terminal shoots into clusters. A webbed cluster of leaves harbour several larvae in the initial stage. The larvae are gregarious and feed by scrapping the leaf surface. Late instar larvae feed individually on whole leaf lamina leaving only midrib. As a consequence of severe feeding, clusters of webbed leaves become dry and colour changes to brown. In case of severe infestation, photosynthetic activity is severely hampered. Affected trees present sickly appearance and can be observed from a distance due to brown dry clustered leaves. Generally, 1-9 larvae are found in a single web. Ninety per cent of the completely defoliated shoots dry and fail to fruit in the forthcoming season. Pupation takes place inside the webs in silken cocoons. However, the last generation (December-January) pupates in the soil. The pupae diapause for about five to six months. The infestation is severe in shady conditions. Old orchards with lesser space between tree canopy have more infestation than open orchards.

Management

- Inspection of mango orchards at least once in a month and searching of for webbed leaves/shoots is essential. Systematically prune the infested clusters and destroy them along with larvae.

Also collect and destroy the webbed infested fallen leaves on the ground under infested trees harbouring old larvae/ pupae. In case of old dense orchards, prune some side branches and open the tree canopy from centre to enable proper sunlight penetration.

- Spray the affected trees after pruning with carbaryl (0.2%) or quinalphos (0.05%) or endosulfan (0.07%). Repeat spray at 10- 12 days interval.

Mites (*Paratetranychus mangiferus, P.yothersi.* and *Eriophyes* sp.)

Among non-insect pests, mites cause damage not only to growing trees but also at seedling stage. Mites, as a mango pest, have been observed in all the mango-growing tracts of India. The mango bud mite, *Aceria mangiferae* Sayed, is one of the most common pests of mango buds. Two types of symptoms are produced by *A. mangiferae,* gall formation and damage to buds. When the buds are attacked and the fungus *Fusarium* sp. is absent, the result is hypertrophy, with a proliferation of stems at the terminal branch. This symptom is known as witches'- broom. When *Fusarium* is present, the trees develop floral and foliar galls. The interaction of *A. mangiferae* with the fungus results in rapid necrosis of hypertrophy, producing multiple budding and subsequent gall formation where flowers and leaves would normally occur. This mite is considered as an efficient vector of pathogens that cause malformation in mango; however, consistent relationship between the mite's presence and symptoms of the disorder has not been found. *P.mangiferus* infests only the upper surface of the mango leaves. The mites spin a profuse webbing and cause mottling of the leaves owing to the disappearance of colour at the spots punctured in feeding.

Management

- Several species of predatory mites, coccinellids and thrips prey upon these mites. For a limited number of trees, removal and destruction of infested parts of plants should be resorted to.
- Spraying the affected plants with lime-sulphur wash (1:2:10) or 0.05 per cent diazinon has proved effective.
- Sulphur, in whatever form it is used, is effective in controlling the mites.

Pests of Mango Trunks, Twigs and Roots

Stem boring Coleoptera , scales and termites are the pests of trunk, twigs and roots.

Termites

Several species of termites have been reported in India which occur in almost all the states. They usually occur in neglected orchards. They also appear and cause havoc in orchards that have been raised in virgin lands. They are active through out the year, though the incidence during the monsoon months is rather low. They either remain under ground to feed fibrous material (cellulose) of roots, shoots and trunks of the mango tree or move upwards making the tunnel completely hollow. Sometimes severely damaged roots dry up. They make earthen galleries on tree trunk and under the safe passage on the trunk they destroy the bark and tissues and branches of the trees. The damage is more severe in nurseries.

Management

- The trunk and branches affected by termites may be sprayed with Chlorpyrifos0.05% after scraping the termite galleries.
- Before planting, remove old stumps and roots from the field and burn them.
- Orchards should be kept clean and free of all refuse vegetation. All the dead and decaying wood, whether on the trees or in the ground, should be regularly removed.
- Application of finely ground *mahua* / *neem* cake, followed by irrigation, helps to drive away the termites.
- Stem of the plants may be pasted with burnt movil oil

Bark Eating Caterpillars *(Inderbela species)*

In addition to mango, bark eating caterpillars attack guava, ber, litchi, orange, pomegranate, *kachnar,* loquat, and mulberry. The moth is light grey in colour with dark brown dots and measures about 35-40 mm with expanded wings. Female lays eggs in batches on the bark. The full grown caterpillar is dirty brown in colour and is about 35-45 mm in length. Larvae actually feed from April to December. There is only one generation in a year.

Mango is damaged to a considerable extent by the bark eating caterpillars, which attack trees of all ages, particularly the older ones, lowering their vitality. The old, shady and neglected orchards are more prone to attack by this pest. When severely infested, the entire branch or tree may die. The attack by this pest is characterized by the presence of long winding, thick, blackish or brownish ribbon-like masses composed of small chips of wood and excreta, both of which intermix with the help of adhesive material secreted by the caterpillar. Larva bores into the bark of the trees and remains in the tunnels dug into the sap-conducting tissues. By continuously devouring the tissues, it tunnels through the stem and branches. This injury weakens the stem, resulting in drying of the branches and finally of the tree itself.

Management

- The caterpillars can be killed by inserting an iron spike into the tunnels.
- This insect has also been successfully controlled by injecting ethylene glycol and kerosene oil in the ratio of 1:3 into the tunnel by means of a syringe and then sealing the opening of the tunnel with mud.
- Another method of control is dipping a small piece of cotton in any of the fumigants, like carbon bisulphide, chlorosal or even petrol, and introducing it into the tunnel and sealing the opening with clay or mud.
- A mixture of equal parts of creosote and chloroform or petrol or cheap kerosene oil may be injected into shelter tunnels to kill the active boring caterpillars.
- Remove the webs from tree trunks and put emulsion of DDVP (0.05%) in each hole and plug them with mud. To control bark eating caterpillar, mix chlorpyrphos 2 ml per litre of water and apply the bark eating caterpillar infested area with a brush at 15 days interval.
- As a preventive measure, spraying of the attacked trunk and branches with 0.05%DDVP may be done.

Stem Borer *(Batocera rufomaculata)*

A wide host range of this pest has been reported including mango. Stem borer is widely distributed in India and attacks a variety of fruit trees including mango. Damage is caused by the grub of this beetle as it feeds inside the stems boring upward resulting in drying of branches and in severe cases stem also dies. However, borer occurrence and injury tends to be sporadic and below levels requiring direct action. Adult beetles, 35-50 mm in size, are stout and greyish brown in colour with dark brown and black spots. Eggs are laid either in the slits of tree trunk or in the cavities in main branches and stems covered with a viscous fluid. Full grown grubs are cream coloured with dark brown head and 90 x 20 mm in size. Pupation takes place within the stem. Beetle emerges in July-August. There is only one generation of the pest in a year.

- Exclude alternative host trees from mango orchards and remove the dead trees and infested branches from the garden to prevent the spread of the pest.
- Swab coal tar + Kerosene (1: 2) on the basal part of the trunk up to 3 feet high after scraping the loose bark in order to deter the females from laying eggs.
- Alternatively, carbaryl 0.1 per cent can be swabbed at bimonthly or put Carbofuron 3G at the rate of 5 g per hole and then plug it with copper oxychloride paste or plug it with mud. Close the holes by the cotton plugs dipped in chloroform or kerosene.
- Apply few drops of Carbon disulfide or fenvularate or Dichlorovos or keep aluminium phosphide tablets in the holes and plug the same.
- If infestation is severe then apply the copper oxychloride paste on the trunk of the tree to prevent disease incidence.

□□□

16

Adopt Preventive Practices and Curative Options for Disease Management

Mango crop is the worst victim of many economically important diseases at all stages of growth and development. And almost all the parts of the plant, *viz.*, root, trunk, branch, twig, leaf, petiole, inflorescence and fruit are invaded by several types of pathogens such as fungi, bacteria, algae, lichen and phanerogamic parasites. Below discussed are most of the important diseases inflicting mango cultivation and their integrated management practices. Integrated disease management (IDM) is an approach that attempts to use all available methods of control of a disease or of all the diseases of a crop plant for best control results but with the least cost and the least damage to the environment. The main goals of IDM program are i) elimination or reduction of the initial pathogenic inoculum, ii) reduction of the effectiveness of initial inoculum, and iii) slowing down the secondary disease cycles. The strategies for managing disease problems may be put into three categories, *viz.*, evasion of the effects of a disease, elimination of susceptibility of crops to attack of diseases (for example, use of resistant varieties), and reduction of inoculum density to levels at which the damage caused is negligible. Translating these principles into tangible management practices will go a long way in reducing the yield losses of mango crop.

Fungal Diseases

Anthracnose (*Colletotrichum gloeosporioides*)

This is the most important fungal disease of mango which is prevalent in every mango growing region. The anthracnose fungus attacks leaves, flowers, twigs and all stages of developing fruit. The severity of the disease is higher during hot humid weather conditions (favourable conditions are high humidity, frequent rains and a temperature of 24-32°C). The disease produces leaf spot, blossom blight, wither tip, twig blight and fruit rot symptoms. This disease is frequently observed on mango nurseries also. Newly emerged leaves are more susceptible to anthracnose. On developing leaves the emergence of small dark brown or black spots is the first stage of infection. With the passage of time, these spots will grow and unite with each other to form irregular shaped brownish black spots. Tender shoots and foliage are easily affected which ultimately cause 'die back' of young branches. When flowers contract anthracnose, both blossom blight and peduncle blight manifest in severe form. The panicles turn black in color and dry off. Severe infection destroys the entire inflorescence resulting in no setting of fruits. Young infected fruits develop black spots, shrivel and drop off. Small developing fruits turn black in color, dried off and fall. Mature fruits show small dark brownish spots at the beginning. As the fruits mature these spots grow in size and turn into black irregular shaped patches. Infected fruits show rotting and crack at the surface with typical 'tear staining' symptom. The disease remains as latent infections in unripe fruit and develops when the mango fruits begin to ripen. Lesions may remain limited to the skin or may invade and darken the flesh. Disease development is encouraged by rains or heavy dews. As the fruits ripen after harvesting, its natural resistance to the disease fades away. Therefore, fungus present on unripe fruit as a latent infection begins to develop rapidly as the fruits ripen. Fruits infected at mature stage carry the fungus into storage and cause considerable loss during storage, transit and marketing. Severely spotted fruits fetch low price in the market.

Management

- Pruning of excess foliage to allow penetration of sun light inside the canopy is essential. Since the fungus has long saprophytic survival ability on dead twigs, leaves and fruits, they should be pruned and burnt along with fallen debris on the orchard floor for reducing the source of inoculum.

- Proper irrigation and balanced nutrition should be provided to maintain good vigour and immunity of the mango trees. Overcrowding of the orchards may be avoided.
- Trees should be sprayed twice with Bavistin (0.1%) at 10 days interval during flowering to control blossom infection. Spraying of Blitox (0.3%) is recommended for the control of foliar infection. Pre-harvest sprays of Captan (0.3%) or Thiophante methyl (0.1%) or Bavistin (0.1%) reduce fruit infections. For effective spray penetration into leaves and flowers, use a nozzle connected to a rubber tube fixed to a wooden pole.
- Post-harvest dipping of fruits in hot water at 52°C for 30 minutes or 15 minutes with Benomyl/Bavistin (0.1%) is effective. Care must be taken to control water temperature and duration of dipping to avoid any damage to fruits.
- Different varieties have varied degree of ability to resist this disease. Maximum damage is observed on Neelum, whereas variety Edward is reported to be resistant.

Powdery Mildew (*Oidium mangiferae*)

Powdery mildew is another serious disease of mango affecting almost all varieties and is prevalent in all mango growing areas of India. Usually it appears from December to March. The fungus attacks leaves, flowers and young fruits during the dry winter. The characteristic symptom of the disease is the whitish, sometimes grayish, superficial powdery fungal growth on inflorescence, tender leaves and young fruits. Generally, the infection starts from the tip of the inflorescence and spreads downwards covering the floral axis, tender leaves and thin stem. Lesions become dark brown and greasy looking as the leaves mature. The affected flowers do not open owing to severe damage, and in most cases shed prematurely. The young fruits may be covered with powdery growth and drop off prematurely, inflicting heavy loss of the crop. The fruits are sometimes malformed and discoloured due to severe mildew attack. The axis also begins to dry, showing characteristic die-back symptoms. Warm humid weather (created by rains or mists) accompanied by cooler nights during flowering are congenial for the disease development and spread of the pathogen within and between trees. The fungus parasitizes young tissues of all parts of the inflorescence, leaves and fruits. Poor fruit set and fruit drop result in very heavy loss of yield, sometimes reaching as high as 70-80 per cent crop loss on individual plant basis.

Management

Though powdery mildew is a very serious disease of mango, fortunately very effective control measure is available with relatively safer fungicides. An integrated approach is presented below as a ready reckoner:

- Regular pruning of diseased leaves and panicles greatly reduces inoculum load of the pathogen.
- Removal of fallen leaves, axis of panicles and malformed fruits from tree base is advised to break the disease cycle of powdery mildew.
- Spraying of Wettable Sulfur 0.2 % (2 g Sulfex / litre water), Dinocap 0.1 % (1 g Karathane/litre water) and Tridemorph 0.1 % (1 ml Calixin/litre water) alternatively is recommended. The spraying should start before flowers open in the inflorescence.
- Neelam, Zardalu, Totapuri, Janardan Pasand, Bangalora have been reported to be resistant against powdery mildew.

Die Back (*Lasiodiplodia theobromae*)

This is a very aggressive and destructive disease of mango making its presence felt in India and other mango growing countries as well. In India, the disease is prevalent in almost all mango growing belts. The disease may be noticed usually round the year on the mango plants but it attains severity during October-November. Characteristic symptoms include drying of twigs and branches from top downward followed by defoliation. In extreme severity, the infected plant takes fire scorch appearance. The bark of infected branches becomes discoloured and eventually dark. The dark area advances and young green twigs start withering at the base affecting the midribs of leaves and then extending outwards along the veins towards leaf edges. The affected leaves turn brown and margins roll upwards. At this stage, the shriveled leaves drop off and the twigs and branches are left bare. This may be accompanied by exudation of gum. Browning of vascular tissue in the affected branch is seen when slit open longitudinally. The areas of cambium and phloem show brown discolouration, and yellow gum like substances are found. In case of infection in the grafting joint, death of the grafted plant usually follows. One of the predisposing factors of the disease is high summer temperature followed by rains.

Management

- Pruning of the diseased twigs and branches, pasting of cut ends with copper oxychloride or Boedeaux paste or cow dung paste and spraying with copper oxychloride (0.3%) can preempt disease invasion. Pruning should be done in such a way that the twigs are removed 2-3 inches below the affected portion.
- Clean orchard management practices should be followed. Orchard floor should not be left with infected mango debris. Maintaining good vigour and vitality of the plants is required through supply of proper irrigation and balanced nutrition.
- Selection of scion from healthy trees, sterilisation of budding knives, placement of budded trees in dry environment and gradual exposure to full sunlight have been suggested for control of dieback of grafted seedlings.

Sooty Mould (*Capnodium mangiferae*)

Sooty mould or black mildew is of common occurrence in India in many kinds of fruits and plantation crops including mango. Though the fungus does not enter the host and derive nutrients from the host tissues, it effectively reduces photosynthetic area of leaves by virtue of its velvety presence. The disease is common in mango orchards where mango hopper, mealy bug, scale insect and coccids, thrive. The disease appears as a black velvety membrane like coating on the leaf surfaces, stems and fruits. In severe cases, the trees turn completely black due to the presence of mould over the entire surface of twigs and leaves. The thin layer formed on the leaf surfaces can be rubbed off easily. The severity of infection depends on the magnitude of honey dew secretion by the above insects. Honey dew secretions from the insects stick to the leaf surface and provide nutritional medium for fungal growth. Mature fruits with sooty mould symptoms do not fetch good price.

Management

- Pruning of affected branches and their prompt destruction prevent the spread of the disease.
- Honey dew secreting insects should be controlled with spraying of appropriate insecticides.
- Spraying of 2 per cent starch is found effective in suppressing the disease.

- It can be effectively controlled by spraying mixture of wettable sulfur + monocrotophos + gum acacia (0.2% + 0.05% + 0.3%).
- Indian oil formulation No.1 and 2 can also be tried at 15 days interval for managing the disease.

Mango Scab (*Elsinoe mangiferae*)

Mango scab can cause significant damage on young seedlings. The fungus causes symptoms on leaves, twigs, barks, flowers and fruits. In early stages, mango scab infection resembles anthracnose, but the spots are smaller in size. The leaf surface is covered with a delicate velvety down. Young succulent leaves show spots on lower surface. In heavy infection, crinkling and distortion of leaves take place followed by premature shedding. Larger spots show shot hole symptoms. The blotches on the stem bark are grayish and irregular in shape. Lesions on fruits are usually grayish brown in colour with dark irregular margins. On mature fruits, spots grow in size with centre of the spot covered with fissured corky tissues.

Management

- Infection in nurseries and new flushes on trees can be prevented by frequent sprays of copper oxychloride (0.3%).
- Spray applications of fungicides for anthracnose control are generally considered effective for scab control.

Root Rot and Damping off (*Rhizoctonia solani*)

This is a serious problem affecting seedlings in nurseries. Sudden dropping of leaves is observed after seedling emergence. Circular or irregular shaped water-soaked patches make girdle at the base of the seedlings. The tissues become soft and dark brown or black. Infection spreads both up and down the seedling base, collapsing it in few days. Presence of fungal mycelia and sclerotia can be seen on dead tissues.

Management

- Soil treatment should be done with formaldehyde 15 days before sowing of root stock stones.
- Biopesticides such as *Trichoderma* should be incorporated in the soil 7 days before sowing.

- Mango stones should also be treated with Thiram or Captain (3 g/litre water) for 5-10 minutes.
- Spraying of copper oxychloride (0.3%) should be done on plants and soil at weekly intervals.

Black Banded Disease (*Rhinocladium corticolum*)

The disease manifests as black velvety raised fungal growth on twigs and branches of mango plants, sometimes on midribs and veins of leaves also. It gradually increases in size and encircles the branches and twigs, thus giving a black banded appearance.

Management

- Commercial cultivars viz., Neelam, Banganapalli (Banishan) and Totapuri (Collector) are susceptible to the disease.
- Application of copper oxychloride or Bordeaux mixture after rubbing the affected portion with gunny bag is very effective in controlling the disease.

Phoma Blight or Leaf Blight (*Phoma glomerata*)

Symptoms are observed mainly on the leaves. Yellowish, pin-head-like spots appear on leaves. The surrounding tissues first become light brown and then dark brown. Initially, the spots are angular, minute, irregular, yellow to light brown, scattered all over leaf lamina. As the spots enlarge their colour changes from brown to cinnamon and they become almost irregular. They have slightly raised and dark purplish margins. The spots become more in number, bigger in size. Later, the tissues become ash coloured due to formation of pycnidia. The round spots gradually become oval or irregular. The leaf tip dries up and infection spreads downwards towards petiole. The lower side of the spot is brown in colour, while upper portion becomes white in centre with a brownish purple margin that separate diseased and healthy portion. In case of severe infection, such spots coalesce forming larger patches and resulting in complete withering and defoliation of infected leaves. The pathogen can survive in mango leaves for over a year. Heavy infection is noticed during the monsoon.

Management

- Since the pathogen survives in leaves, removal and destruction of infected plant parts from the orchard can reduce the amount of inoculum.

- The disease can be kept under control by spraying benomyl (0.2%) followed by copper oxychloride (0.3%) just after the appearance of the disease and subsequent sprays at 20 days intervals.

Pink Disease (*Erythricium salmonicolor*)

The disease is widespread on many fruits and plantation crops not only in India but in many tropical and sub-tropical countries. The disease appears as a pinkish, powdery coating on the stem bark. The pink colour represents profuse conidial production by the fungus. The fungal growth may spread and girdle the stem. The pathogen invades cortical tissues and interferes with transport of nutrients. In advanced stage of disease development, infected bark gets shredded baring the wood. Shoots and branches of the affected trees wilt and dry up, and the leaves shed away.

Management

- Removing the infected branches and pasting them with Bordeaux paste keep the disease under control.
- Canopy management to allow air circulation and light penetration can be helpful.
- Painting or spraying of the affected part by lime sulfur or copper hydroxide or tridemorph is effective for controlling the disease.

Grey Blight (*Pestalotiopsis mangiferae*)

Brown spots appear on the leaf lamina, originating from margins of tip. Small light brown spots gradually increase in size, become dark brown and coalesce to form large patches on the leaf lamina. When the infection starts from tip, both sides of the midrib are engulfed by infection and finally get detached from the leaf. Brown spots also appear on fruits, turning grayish white at later stage.

Management

- Chausa is a resistant cultivar to the disease while Langra is least susceptible.
- Spraying of wettable sulfur (0.2%) before and after flowering, Bavistin (0.2%) at pea stage, and Zineb (0.2%) before maturation of stone reduces the incidence of grey blight.

Gummosis (*Lasiodiplodia theobromae*)

Gummosis of mango is slowly becoming a serious problem in mango growing areas of India. The disease occurs on stem and branches of the plants. The disease makes its appearance in dry seasons, mainly in winters. Profuse oozing of gummy substances on the wood surface and bark of the cracked branches is the characteristic symptom of the disease. With increasing severity, the gum drops may roll down the stem bark, setting off rotting of the bark.

Management

- Removal of diseased barks and pasting with Bordeaux paste or cow dung paste help in keeping the disease under control.
- Application of copper sulfate @ 500 g in soil around the tree base is a recommended practice for management of the disease.

Alternaria Spot (*Alternaria tenuissima*)

Alternaria spot appears on leaf and fruits. Minute circular spots are seen on lower surface of leaf and on fruit. With passage of time, the disease is visible on upper surface of leaf in the form of irregular large patches. Latent infection on the fruits start showing symptoms during ripening and coalesced spots form sunken patches on the ripe fruits.

Management

- As fallen leaves on the ground are the sources of infection, burning them will reduce chances of disease incidence.
- Postharvest treatment of fruits with Procloraz gives effective control of the disease.

Ganoderma Root Rot (*Ganoderma lucidum*)

Ganoderma is a soil-dweller, invading roots and spreading upwards to the basal portion of the stem. The stem becomes soft and gradually degrades. The fungus spreads rapidly and forms bracket like fruiting bodies on the side of the stem. In extreme situation, the mango plants may wilt. Infected roots and stems lose weight and get crumbled very easily. The fungus also attacks other fruit trees like citrus, coconut, cashew etc.

Management

- When the disease is noticed on one or two trees, the area around the plant should be isolated by trenching and the infected plant should be uprooted and burnt.
- The healthy plants should be given protective coating on the trunk with Bordeaux paste and the soil around the plant should be drenched with Bordeaux mixture.
- Organic manures amended with *Trichoderma* should be applied in the soil heavily.
- Fruiting bodies (brackets) of the fungus should be manually collected and burnt.

Verticillium Wilt (*Verticillium albo-atrum*)

Verticillium wilt can occur in the limestone soils and is usually observed in new trees planted on land previously used for vegetable production (especially tomatoes). This fungus attacks the tree roots and vascular (water conducting) system, decreasing and blocking water movement into the tree. Symptoms of infection include leaf wilting, desiccation and browning, stem and limb dieback, and browning of the vascular tissues. Verticillium wilt sometimes kills young trees partially ('one sided' wilt).

Management

- Soil drenching with Bavistin (0.1%) or Captafol (0.25%) has been found effective to control the disease.
- If possible, areas with intensive vegetable cultivation should be avoided for establishment of new mango orchards.

Sclerotium Rot (*Sclerotium rolfsii*)

The disease infects stones and emerging seedlings, causing extensive rotting of stones and stems of seedlings. Characteristic symptom of the disease is the development of weft by fungal mycelia on stem base near the soil level. Dark brown spots may also be seen beneath the weft. The seedlings droop and bend towards ground, culminating in death within a week.

Management

- Soil in the nursery bed should be surface burnt to disinfect before sowing of mango stones.
- Need based watering should be resorted to and proper spacing of stones must be followed. Adequate drainage arrangement is to be provided.
- Diseased seedlings and adjacent weed hosts should be removed from the vicinity of the nursery.
- Stones should be treated with Thiram or Captan (0.2%) before sowing. Intermittent soil drenching of nursery bed with same fungicides should be followed.

Stem End Rot (*Diplodia natalensis*)

It is an important postharvest disease of ripe mango fruits. The rot not only spoils 4 to 6 per cent fruits every year in the Indian markets but also affects the mango export. Disease symptoms develop around the stem end of the fruit as it begins to ripen after harvesting. In the initial stage, the epicarp darkens around the base of the pedicel. At the stem end of the fruit, brownish patches begin to develop. In the next few hours, the affected area enlarges to form a circular, black patch which, under humid atmosphere extends rapidly and turns the whole fruit black within two to three days. Infected fruit gives a bad smell, is brown in colour and somewhat softer. The infected fruits lose ascorbic acid rapidly and non-reducing sugars are low in the pulp of such fruits. Dead twigs and bark of mango trees harbour the pathogen abundantly, and with the onset of rains, the microclimate of the mango orchard gets contaminated with spores of the fungus. The fungi may form latent infection on fruit.

Management

- Harvesting of fruits should be done on a clear dry day. Fruits should be covered and removed to the ripening houses immediately after harvest. Care should be taken to prevent snapping off of the pedicel. If the fruits are harvested without stalk, wax should be applied to the snapped stalk ends immediately. Injuries to the fruits should be avoided at all stages of handling.
- In areas where this disease is severe, a spray schedule with Benlate (0.3%) or Bavistin (0.1%) one month before harvesting is advised. Sprays must be repeated in 15 days intervals.

- Clean cultivation practices should be followed. Leaves and twigs should not be allowed to fall and rot near the trees. Avoid harvesting of fruits that are not sufficiently mature. Store harvested fruits at low temperature.
- Hot water treatment of fruits for 5-10 minutes (depending on fruit size) in water having 50°C ± 2°C (122°F ± 4°F) temperature and 0.5% Bavistin is strongly recommended.

Black Rot (*Aspergillus niger*)

The affected fruits show yellowing with irregular, dull grayish spots which coalesce to form dark brown or black necrotic areas. The black mould is found to grow on the infected fruits. The mesocarp of the rotten area becomes depressed and soft. Disintegrated tissues of the fruit emanate bad smell. The stalk-end infection results in premature fruit drop.

Management

- Dipping of fruits with Benlate @ 0.1% can control the mould fungus.
- Hot water treatment of fruits for 5-10 minutes (depending on fruit size) in water having 50°C ± 2°C (122°F ± 4°F) temperature and 0.5% Bavistin is strongly recommended. This is effective against stem end rot also.
- Injury of fruits should be avoided. Care should be taken not to allow fruits to touch the soil.

Bacterial Disease

Bacterial Canker (*Xanthomonas campestris* pv. *mangiferaeindicae*)

Bacterial canker of mango (also known as mango bacterial black spot or MBBS) is rapidly attaining threatening proportion in all mango growing areas of India, causing as high as 100% loss in some varieties. Besides being pathogenic on several varieties of mango, the organism is capable of infecting wild mango, cashew nut and weeds as well. Many commercial cultivars of mango including Langra, Dashehari, Amrapali, Mallika, Totapuri, Arka Puneet, Arka Neelkiran, etc. are susceptible to this disease. The disease makes its appearance on leaves, petioles, twigs, branches and fruits. The disease first appears as minute water-soaked irregular lesions on leaf bound by the veins. Initially, the lesions are light yellow in colour, but as the disease advances the lesions enlarge and turn dark brown surrounded by chlorotic haloes. With age, they become angular, cankerous with black raised growth. Several lesions coalesce to form irregular necrotic cankerous patches. In severe infections, the leaves turn yellow and drop off. Cankerous lesions appear on petioles, twigs and young fruits. The water soaked lesions develop on fruits which later turn dark brown to black, gradually developing into cankerous raised spots. The spots often burst open, causing cracking, splitting and releasing highly contagious gummy ooze containing bacterial cells. Mango pulp and stones are also reported to be affected in severely infected fruits. The fresh lesions on branches and twigs are water soaked which later become raised and dark brown in colour with longitudinal cracks but usually without any ooze.

Management

- Seedling certification, regular inspection of mango orchards and maintaining good sanitation are essential to prevent the disease. The disease can be evaded by using disease-free rootstock and scion obtained by spraying with copper oxychloride (3 g/litre). Scions may also be disinfected with 0.35% calcium hypochlorite or sodium hypochlorite solution for 5 minutes before grafting. Grafted plants should be raised away from sources of infection by the disease.
- Use of wind breaks prevents mango plants from bruising or wounding, thus reducing chances of infection. Preventative sprays of copper oxychloride @ 3 g/1 litre helps to evade the disease.
- Three sprays of streptocycline (200 ppm) or Agrimycin-100 (100 ppm) after first visual symptom at 10-days intervals gives good control of the disease.
- Streptocycline (300 ppm) and Copper oxychloride (0.3%) can also be sprayed alternatively for control of the disease.
- One should not wait for appearance of the disease. The key to success for complete control of the disease is to start spraying before the infection has established.

Algal Diseases

Red Rust or Algal Spot (*Cephaleuros virescens*)

Red rust disease, caused by an alga, has been observed in almost all mango growing areas. This parasitic alga attacks leaves and stems.

The disease is more common in closely planted orchards. Fruiting bodies of the alga are formed during humid atmosphere. The zoospores formed by the sporangia initiate fresh infections. The algal attack causes reduction in photosynthetic activity and defoliation of leaves thereby lowering vitality of the host plant. The disease can easily be identified by the rusty red velvety growth (fructifications) mainly on leaves and sometimes on petioles and bark of young twigs. Later, they turn reddish brown. The circular and slightly elevated spots sometimes coalesce to form larger and irregular spots. The alga is epiphytic in nature. Stem entry is achieved through crevices. The affected areas crack and scale off, and can lead to bark cankers and stem death.

Management

- Nutritional deficiency should not be allowed to occur in the mango plants. Close planting should be avoided in high humidity areas. Proper canopy management must be adhered to for keeping the disease away.
- Two to three sprays of copper oxychloride (0.3%) during July at 15 days interval is effective in controlling the disease.
- Algaecides such as Fentin acetate have also been found effective to control the disease.

Phanerogamic Parasites

Mango tree falls prey to some phanerogamic parasites and epiphytes during its course of life. The parasites are found parasitizing on the trunk or branches of the tree drawing nutrition from the host.

Dendrophthoe (earlier name *Loranthus*)

Dendrophthoe (*Loranthus*) is more commonly involved as a parasite in mango than any other phanerogamic parasites. *Dendrophthoe falcata* is the most destructive parasitic pest of mango in India. In North India, more than 60 per cent of mango trees are affected by this stem parasite, especially in old and neglected orchards. The parasite flowers and bears fruits, and its leaves are more or less similar to that of mango with lesser length. Birds eat the fruits of the parasite and the seeds are excreted through droppings which fall on limbs and branches of the mango trees. The seeds germinate under favourable conditions and give out clusters of shoots. The seeds while sprouting produce bulged haustoria which penetrate into the bark and absorb water and minerals from the host, and the parasite grows as an individual plant at the cost of the mango tree. The point of entry into the host tissue by the parasite is characterized by swollen growths called 'burrs'. The haustoria of the parasite serve as roots. Severely attacked trees produce small leaves of unhealthy green colour. The quality and yield of fruits from such trees are considerably lowered.

Management

- The parasite can be easily detached from the host mango plants in the early stages.
- In the later stage, the infected branches should be cut to eradicate the haustoria. Application of Bordeaux paste to cut ends of the tree is to be done without fail.
- Spraying of 30-40 per cent diesel oil emulsion in soap water or injection of copper sulfate to the infected branches may help eradicate the parasite.

❑❑❑

17

Adopt Preventive Practices and Curative Options for the Management of Disorders

Various kinds of disorders regularly afflict mango crop. These disorders may be due to biological agents, environmental factors, cultural conditions or some other reasons. They are primarily the results of imbalances in metabolism of the plant induced by some biological and/or physical factors.

Mango Malformation

Mango suffers a colossal loss due to malformation, a serious threat to mango cultivation. This is a complex malady and is one of the most destructive disorders. The economic losses faced each year due to malformation vary between 5-30 per cent. Malformation is the production of thick vegetative shoots at the growing tip and transformation of floral parts into a compact mass of sterile flowers. It has been reported from almost all the mango growing regions of India. It is widely prevalent in the north while its incidence is rather sporadic in the south. Two types of malformation have been observed, viz., vegetative and floral.

Vegetative malformation : The vegetative malformation is more pronounced in seedlings and young plants. The affected plants develop excessive vegetative branches of limited growth. The numerous small shootlets which arise from apex of shoot or leaf axil get thickened and clustered together, giving the twig a bunchy top appearance. The axillary buds on these branches are usually swollen. Similarly, the affected branches of older plants develop swollen axillary buds but the growth of these branches is not impaired. The malformed bunches dry up after a few months but the branches or twigs bearing them continue to grow in spite of the disorder. Malformed seedlings usually have shallower root system with fewer tertiary roots. It is generally observed that trees between 4 to 8 years of age suffer the most from vegetative malformation.

Floral malformation: In this type of malformation, abnormal inflorescence is developed in which individual flowers in a panicle are bigger with large discs, and the peduncle bearing the flower is fairly thickened. Compact masses of mostly sterile flowers appear in the form of a bunch. A few malformed flowers are hermaphrodite. Malformed panicles carry more number of flowers than the normal. Three groups of malformed panicles can be differentiated on the basis of compactness: heavy, medium and light. With passage of time, part of inflorescence becomes vegetative giving rise to small leafy structures.

Cause of malformation : The exact cause of this monumental problem of mango production is still a matter of controversy. There are many hypotheses doing the rounds to explain the cause of the problem. Physiological disorder, eriophyid mite (*Aceria mangiferae*) and virus were earlier suspected to be the cause of the disease. It is now considered to be caused by *Fusarium moniliforme* var. *subglutinans*. The fungus has been isolated from swollen axillary buds, infected twigs, inflorescence axis and flowers. The advance of the disease can be checked by pruning the infected twigs. A hypothesis is also proposed that the fungus produces malformin and causes malformation.

Management

- Many researchers suggested regular pruning of malformed parts (shoots and inflorescence) for checking of mango malformation. Systematic pruning and burning of affected parts result in gradual decrease in incidence of the malady. Pruning followed by spraying with the mixture of a fungicide (captan 0.1 %), acaricide (Akar-338 0.1 %) along with a sticker is found effective against malformation.

- Mango plants should be provided with balanced nutrition to maintain good health and vigour. The orchards may be inspected quite frequently and any malady noticed may be removed from time to time. It is suggested that the orchards may be kept in good hygienic conditions and disease free planting material should only be used for planting.

- Spray of NAA (100 to 200 ppm) during the first week of October has been worked out to reduce floral malformation. Later, four sprays of 250 ppm NAA at weekly intervals from 20th October onwards have been suggested to reduce the intensity of malformation to the maximum possible. Spraying of 1000 ppm paclobutrazol (10-60 g/tree) during the first week of October reduces malformation.

- Deblossoming at the bud burst stage alone or in combination with the spraying of 200 ppm NAA was found effective in management of malformation.

- Anti-malformin has been observed to reduce the activity of malformin in the mango plant, thus reducing the extent of malformation. Glutathione @ 2240 ppm and ascorbic acid @ 2110 ppm were also found to be effective in reducing the severity of malformation. The chemicals may be applied three times after appearance of malformation.

- The application of 0.04% diazinon was found to control malformation. Two sprays of 0.32% diazinon could minimize malformation. It was reported that 99% control of the disease was attained by excising the desired shoots 300-600 mm below the inflorescence, burning them and spraying trees with 0.1% parathion, 0.1% thiometon after every three weeks. The simple removal of shoots in July and August had given 98% control with no disease for next two years.

- Polythylene film covers used around mango shoot in winter controlled malformation. It was suggested that malformation could be controlled by raising temperature around the panicles or orchards in winter. The temperature could be raised by covering mango tree with polythene tent before the onset of flowering.

- The most simple, effective, economically viable and eco-friendly approach that has been tested at IARI, New Delhi (reported in The Hindu dated 19th August, 2004) to control this malady is the use of the leaf extracts of *Datura stramonium, Calotropis gigantea* and neem tree (*Azadirachta indica*). Slurry is prepared by mixing

100g of cow manure and crushed leaves of *Datura*, neem and *Calotropis* (ca. 300g) in 50 litres of water. It is allowed to ferment for 20-30 days with occasional stirring. The extract is then filtered and sprayed on mango trees. The malformed panicles when sprayed with the above extract dry up within two days, but remain hanging on the tree. These dried panicles are removed during pruning operation to reduce inoculum levels.

Black Tip

The first outward symptom is the development of a small, etiolated area at the distal end of the fruit which spreads and covers the whole tip. The damage to the fruit gets initiated right at marble stage with a characteristic yellowing of tissues at distal end. Gradually the colour intensifies into brown and finally black. At this stage, further growth and development of the fruit is retarded and black ring at the tip extends towards the upper part of the fruit. Black tip disorder has generally been detected in orchards located in the vicinity of brick kilns. It has been reported that the gases like carbon monoxide, sulfur dioxide and ethylene constituting the fumes of brick kiln are known to damage growing tip of fruits and give rise to the symptoms of black tip. A deficient supply of boron resulting from the interaction of constituents of brick kiln gases with cell metabolites of mango leaves and fruits may cause the disorder. The affected fruits become unmarketable and reduce the yield to a considerable extent. Apart from these factors, climatic conditions, tree vigour, irrigation, and crop management practices also play important role in deciding the severity of the disorder.

Management

- The incidence of black tip can be reduced by the spray of boron in the form of borax (0.6%) or other alkaline solutions like caustic soda (0.8%) or washing soda (0.5%). Four sprays of borax at flower bud stage, full bloom stage, fruit setting stage and pea size stage are recommended.

- For prevention of black tip disorder, the following steps should be taken: i) establishing of mango orchards in North-South direction and 5 to 6 km away from the brick kilns, ii) restricting new kiln sites at a safe distance (1-2 km), iii) regulating the working of kilns before and at the time of fruit-set, iv) making the kilns smokeless or use of high altitude chimneys (more than 50 ft high).

Soft Nose

Mango soft nose is defined as the breakdown of flesh on ventral side and towards the apex of the fruit prior to harvest found mostly in Kent cultivar grown in Florida. Soft nose is found to occur in fruits which are left to ripen on the tree. The malady is severe in late maturing variety particularly in delayed harvest leading to substantial loss. Softening of tissue starts at the apex. Flesh appears over-ripe and may discolor and become spongy. The affected tissue is visible after the ripe fruit is sliced. Hard tissue is present adjacent to soft tissue. Numerous soft nosed fruits are seen on tree during fag end of summer with onset of showers. The fruits with this disorder are unfit for export. This disorder is different from spongy tissue primarily occurring in Alphonso mango of India in both external and internal symptoms of the affected fruits.

Management

It has been found that high nitrogen level in the tree aggravates the disorder while high calcium level mitigates the problem. Hence, the management strategy should be:

- Supplying balanced nutrition to the plants for maintaining nitrogen in just optimum level.
- Maintaining optimum calcium level of the mango plants as well as soil.
- Harvesting the fruits at early stage to avoid the disorder.

Internal Necrosis

This disorder is mostly found in immature fruits on the tree. Affected fruits drop off before attaining maturity inflicting loss to crop yield, particularly when mango is grown in sandy soil. The typical symptoms characterize one or all of the following: i) a softening (breakdown) and water soaking of the fruit flesh at the distal end while the flesh around the shoulder remains healthy, ii) an open gummy cavity in the pulp at the stem end, iii) over-ripe flesh next to the seed surrounded by relatively firm flesh, iv) areas of varying size in the flesh appearing spongy with a grayish-black color. Internal necrosis is supposed to be the result of boron deficiency in the soil, unlike black tip which is caused by deficient supply of boron resulting from the interaction of constituents of brick kiln gases with cell metabolites of mango leaves and fruits.

Management

- Application of borax directly in the soil @ 500 g per tree or foliar spray of the same @ 1% alleviates the problem of internal necrosis.
- This disorder is also aggravated by high nitrogen fertilization. Balanced fertilization should be adopted.

Clustering of Fruits

Clustering is a fruiting disorder in mango, locally known as 'Jhumka'. It is characterized by the development of fruitlets in clusters at the tip of the panicles. Such fruits do not grow beyond pea or marble stage and drop down after a month or so of fruit set. These fruits do not contain seeds when they are cut open. The disorder seems to be due to lack of pollination/fertilization which may be attributed to many reasons. Among them, absence of sufficient population of pollinators in the orchards is the major reason. Surveys conducted in the mango belt of Lucknow revealed that the problem is more severe in areas around Malihabad. Dashehari mango is the most affected cultivar. The other reasons causing the disorder are old and overcrowding of trees, indiscriminate spraying against pests and diseases, use of synthetic pyrethroids for spraying, monoculture of Dashehari and aberrant weather conditions during flowering.

Management

- Spraying of insecticides should be avoided at full bloom stage to avoid killing of pollinators.
- Pests and diseases should be controlled in time by spraying the recommended pesticides only.
- Introduction of beehives in the orchards may be resorted to during flowering season for increasing the number of pollinators.
- The practice of monoculture of a particular variety may be avoided. In case of Dashehari, 5-6 per cent of other varieties should be planted in new plantations. In old orchards, where monoculture of a particular variety like Dashehari is followed, a few branches may be top worked with pollinating varieties.
- Pruning of old trees may be done to open the canopy.
- Spraying of 300 ppm NAA may be done during October-November.

Spongy Tissue

It is a major problem in Alphonso, where a pulp patch fails to ripen. In addition to Alphonso, few other mango cultivars such as Jamadar, Vanraj, Olour, Goamankur, Vellaikolamban, Swarnarekha and Fernandin are also reported to attract the problem. The peculiarity of this malady is that the external symptoms of spongy tissue affected fruits are not apparent either at the time of picking or at the ripe stage. The affected tissue is visible only when the ripe fruit is cut into two halves. The affected fruits become unpalatable and give offensive smell. This malady may be caused by the inactivity of ripening enzymes due to high temperature, convective heat and post harvest exposure to sunlight. As a result of this malady, the quality of affected fruit is very much impaired and the fruits fetch low price in the market.

Management

No established method of controlling spongy tissue is available till date. However, preventive practices may reduce the disorder in susceptible cultivars, especially Alphonso.

- Use of organic mulches with dry mango leaves and paddy straw.
- Practicing sod culture having natural vegetation with *Desmostachya binnata* (dhorth).
- Harvesting of fruits at 3/4 maturity stage to obfuscate the disorder.
- Post-harvest exposure of fruits to low temperatures (10-15 °C) for 10-18 hours.

Leaf Scorching

Leaf scorching is a nutritional disorder in mango. It involves collapse of leaf tissues in very clearly defined marginal areas of brick red colour primarily towards the tip. In severe cases, the whole leaf may be scorched. The symptoms are very distinct and can be spotted out easily from a distance on the tree. The disorder has been found to be the expression of excessive accumulation of chloride ions in the leaf tissues.

Management

- Chloride containing fertilizers should not be used. Muriate of potash should be replaced with potassium sulfate.

- Irrigation water containing high chloride ions should be avoided.
- Fallen scorched leaves should be removed and destroyed from base of the plant.
- Foliar application of potassium sulfate (5%) on new flushes helps to mitigate the disorder.

Fruit Drop

Despite initial high fruit set, the ultimate retention of fruits is quite low in mango. The intensity of fruit drop varies from variety to variety. The fruit drop can be classified into three groups: i) pinhead drop, (ii) post-setting drop and (iii) May-month drop. The fruit drop in first two groups is insignificant compared to the third group which affects the final yield significantly and needs attention. Deficient nutrition of many developing embryos may be the most important internal factor leading to post-fertilization fruit drop in mango. This results in due to competition among over-crowded fruitlets on a panicle. Degeneration of the embryo in the initial stages of its development may yet be another cause of fruit drop. This occurs invariably even if the flowers are self-pollinated. Other factors include climatic factors, disturbed water-relation, disease, pest and hormonal imbalance.

Management

Except in some varieties such as 'Langra' and 'Chausa', natural fruit drop does hardly pose a problem in normal bearing. However, the following practices may be adopted to overcome the problem of fruit drop.

- Drop due to insect and disease attack must be checked in all the varieties by applying appropriate insecticides and fungicides.
- Cross-pollination of the varieties ought to be assured by avoiding isolated planting of a single variety.
- Balanced nutrition of the tree ensures proper embryo development and rectify in many cases hormonal imbalance.
- Several growth-regulators are tried to control the fruit drop. The foliar application of Alary (B-nine) @ 100 ppm or NAA @ 200 ppm at pea stage of fruit was found effective in controlling fruit drop in mango.

Stem End Cavity

The physiological disorder known as 'stem end cavity' has not been adequately described or recognized internationally as a separate disorder, and as such is easily confused with terms such as sunburn, stem end rot, jelly seed and soft nose, etc. Vascular fibres that connect the mango stem to the flesh and top of the seed snap or tear thereby forming a cavity within the flesh under the stem. This results in breaking of sap pressure in the fruit and premature ripening of the flesh around the tear and seed. Conditions favourable to the disorder include cultivar, high leaf nitrogen, low leaf calcium levels, wet windy conditions and harvesting of tree ripe fruit. The first signs of the disorder are a sunken grey/brown area opposite the suture near the stem attachment. Affected fruit suffer premature fruit ripening followed by extensive wind drop. External symptoms are not always evident.

Management

- Early harvesting of the fruits help in reducing the disorder.
- Discarding of all fruits that fail to exude sap in the desapping operation.
- Balanced nutrition of the mango trees helps them to fight against the disorder.

Sun Scald (Heat Injury)

The mango fruits, exposed to direct summer sun or excessive heat during hot water treatment, get damaged. In course of time the rind becomes hard with dull colour. Thus, the fruits become unfit for consumption. Exposure of fruits to temperatures above 30°C (86°F) for longer than 10 days results in uneven ripening, mottled skin and strong flavour. Exceeding the time and/or temperature combinations recommended such as 46.4°C (115.5°F) water dip for 65-90 minutes (depending on fruit size) causes heat injury (skin scald, blotchy coloration, uneven ripening).

Management

To reduce the incidence of scald damage to fruits, the following procedures may be followed:

- Irrigation should be reduced or cut off at least two weeks before harvest.

- Harvesting of only mature fruits should be done for hot water treatment.
- Oils should not be used for desapping.
- Hot water treatment should be delayed for up to 24 hours after harvest.
- The dip tank for hot water treatment should have good circulatory system.
- The water temperature and dipping period must not exceed the prescribed limit.

Chilling Injury

Storage of mango fruits for longer periods below a critical temperature level causes chilling injury. The chilling injury manifests as definite pitting on the surface, darkening and softening of the tissues. The chilling injury causes leakage of metabolites such as amino acids, sugars and minerals from the cells. Symptoms include uneven ripening, poor colour and flavour, surface pitting, grayish scald-like skin discoloration, increased susceptibility to decay and flesh browning. Chilling injury incidence and severity depend on cultivar, ripening stage (ripe mangoes are less susceptible), temperature and duration of exposure.

Management

- 4% calcium chloride impregnated at 350 mm Hg or 200 mm Hg subatmospheric pressure is found effective in controlling the chilling injury in Baneshan mangoes upto 3 weeks.
- Storage time below critical temperature level should be adjusted in such a way that it does not inflict chilling injury.

Bark Cracking

The malady manifests as longitudinal cracks on trunk and limbs. Bark dries in advanced stage. The leaves in the affected tree are sparse, pale and lustreless. The tree exhibits wilting symptoms in extreme cases, and with passage of time the entire tree dries up. Most conspicuous symptoms are rotting of roots and adherence of dried leaves to twig.

Management

- One spray with copper oxychloride 3g/litre or application of 250 g copper sulfate per tree basin contains the malady to a great extent.
- Pasting of tree trunk with Bordeaux paste/cow dung slurry also reduces the problem.

Woody Gall

Woody galls of 10-15 inches diameter are formed on limbs and branches. The galls are abundant on Chinnasuvarnarekha, Langra and moderate on Neelam and Totapuri. Galls affect the health of the tree.

Management

- Remedy lies in removal of galls using saw and applying Bordeaux paste to cut surface.

Sap Burn

When mangoes are harvested, the sap is spilled out. If the sap touches the fruit, it causes blemish. This is called sap burn. If crops are damaged during or after harvest, they get spoilt quickly. Sap burn reduces price of the mango fruits in the market.

Management

- At the time of harvesting, length of stem is left attached to the fruit. The mango is held upside down and the stem is clipped off very close to the fruit. It's important to hold the mango upside down, so that the sap will spurt or drip downwards when the stem is clipped. The fruits are placed upside down for about 30 minutes until the sap has drained out of the stem fully.

Scarified Fruit

The fruit surface constantly touching the ground or branch or leaves develops elephant skin like discoloured hardy rind due to tissue damage and healing. The quality of the fruit is reduced substantially.

Management

- To prevent scarification, branch of the fruits touching the ground should be raised by giving support.

- Zigzag branches should be removed from the tree to avoid scarification of fruits.

Skin Abrasions

Abrasions due to fruit rubbing against rough surfaces or each other result in skin discoloration and accelerated water loss.

Management

- Careful handling of the fruits during harvest, transport and storage reduces skin abrasions.

Lenticels Spotting

This may occur during the pre-harvest or post-harvest period. In pre-harvest spotting, when developing and maturing mango fruits are exposed to prolonged wet periods (especially during rainy season for late maturing varieties), these fruits exhibit the black needle spots at harvest, which become more apparent as the fruits ripen. In case of post-harvest spotting, this malady becomes evident when mango fruits are washed and soaked in water with detergent for more than five minutes.

Management

- Prolonged wet periods for the fruits should be avoided.
- Harvested fruits should not be soaked in detergent water for washing.

Little Leaf Due to Zinc Deficiency

Terminal flushes on the upper portion of the trees show first symptoms. Leaf blade gets thickened and does not attain normal size. The margins of leaf may bend downward or upward. Leaf veins develop yellowing colouration. Leaf blade becomes brittle; internodes get shortened. Crowding of leaves at twig apex gives rosette appearance.

Management

- Spraying of 1% zinc sulfate or 0.2% zinc oxide is recommended.

Copper Deficiency

Young trees show copper deficiency by putting forth weak terminal shoots followed by defoliation and die back of the branches.

Management

- The deficiency can be easily mitigated by application of copper fungicides. In fact, the copper deficiency symptoms are rarely observed in mango orchards where copper fungicides are sprayed for control of leaf diseases.

Frost Injury

In colder areas, mango trees are affected by frost injury. Young mango plants are more prone to frost injury in December-January. Gum exudation from split stem bark, death of new shoots, charred appearance of leaves and inflorescence are the symptoms of frost injury.

Management

- As the plants die immediately due to frost injury, the plants should be protected by thatching, and keeping the east side open for sunlight to penetrate.
- Irrigating the plants by overhead irrigation system is also helpful.
- Pruning of affected twigs and branches is advised in case of partial death of plants.

Jelly Seed

In jelly seed disorder, the flesh around stone gets disintegrated into a jelly-like mass. Loosening of pulp, formation of jelly, unpleasant flavour and pale colour of flesh around stone are characteristic symptoms. Late maturing pulpy cultivars such as Dashehari, Amrapali and Mallika are more prone to this disorder.

Management

- Recently, it has been reported (ICAR News, April-June, 2007) that maximum incidence (>60%) of jelly seed was noticed in Dashehari fruits harvested after second week of June or after a spell of heavy rains. For managing this disorder, an integrated approach was found very effective which involved application of black plastic mulch (100 μ thick) during October-November, foliar spray of dehydrated calcium chloride (2.0%) and potassium sulfate (1.0%) one month before harvesting of fruit and application of 250 g borax per tree in soil during November.

Black Pimples on Fruit

Small, black, raised spots of pin head size are formed on fruit. These spots neither enlarge nor rot, but degrade the quality thereby loosing consumer preference.

Fruit Tumors

Sometimes, tumors of pea to marble size develop on fruits and they render the fruits very ugly look. The stylar end part is more vulnerable than stem end portion. The tumors are seen abundantly on Totapuri. The etiology is not known. Fruits with tumors have least consumer preference. They are neither fit for export nor for pulp making in canning factories.

□□□

18

Adopt Monthly Work Schedule to keep Mango Orchard Healthy and Productive

It has been observed that most of the mango growers are active in their orchards only when flowers or fruits are present on the trees. They have least concern about their orchard for rest of the period. This practice is not good and leads to many problems and poor productivity of the orchard. To maintain vigour, growth and productivity, it is necessary to perform proper operations throughout the year. In the following text suggestions / recommendations based on research work are given regarding various operations required to be carried out during different months.

January

Care of the newly planted saplings and small plants with particular attention to irrigation and cold management should be done. Looking to the previous pest record, arrangements to procure pesticides for the control of mango hoppers, thrips, powdery mildew, anthracnose, blossom blight etc may be done. Moisture management may be done carefully as bud burst starts from November to January in mango depending upon the area of the mango orchard. The polythene band which is applied in the month of December to prevent ascent of nymphs of mealy bug becomes dusty. This should be cleaned.

Powdery mildew is very harmful to mango since it destroys inflorescence. Remove malformed panicles and leaves, and burn them. Spraying of wettable Sulphur (2 g/litre of water) is quite useful to control it. Mix a liquid soap with the fungicide as it helps stick well on plant surfaces. About 10-20 litres of solution is enough for a fully grown plant. It may be increased according to the canopy size of the plant.

In certain cases the inflorescence are elongated to about 3 inches during this month and hopper become very active. To control the hopper the method mentioned in this book may be followed. These sprayings may be done along with the spraying of the insecticides as suggested for the control of hoppers. This means in all the three sprays one insecticide and one fungicide may be mixed and sprayed. Mixing of insecticide and fungicide with neem products may be avoided. Similarly mixing of insecticides with copper oxychloride be also avoided.

Febraury

Spraying against mango hoppers/flower midge and powdery mildew disease to be continued. Removal of thatches erected over the young plants for protection against frost may be done in the last week. Care of intercrops and sowing of summer crops like Okra and cucurbetaceous crops may be done. Areas where fruit borer and stone weevils are problem, the timely action may be taken. Powdery mildew continues to play havoc. Spray of Tridemorph or Calixin (1 ml/litre of water) should be done to increase percentage of fruit set. One spray of Dinocap or Tridemefon should also be done after fruit set. Mix liquid soap with this fungicide for better sticking ability and effectiveness. On initiation of black spots on panicles caused by blossom blight or anthracnose, spray Carbendazim or Bavistin (1 g/litre water). Remove affected leaves and twigs and burn them. It keeps the inoculum load of disease under control.

March

There may be need of spraying against powdery mildew and mango hoppers which may be done as per schedule. When the fruit is of pea size spray of nepthalene acetic acid 20 ppm (2 ml chemical in 4.5 litres of water) is recommended. This is done to prevent fruit drop. To control black tip disorder or internal necrosis, spray of 1% borax (10 gm chemical/litre of water) may be done when fruits are of pea size. Three sprays may be done at 15 days intervals. Removal of thatches erected over the young plants for protection against frost may be done in the first week, if not done earlier. Irrigation may be done at 15 days interval for 2-3 times when the fruits are of pea size. Pruning and burning of hanging malformed panicles should be carried out. Spray for the control of fruit borer and stone weevil may be done.

In any possibility of powdery mildew, repeat spray of Dinocap or Tridemefon should be done. Second spray of Carbendazim should also be done to keep anthracnose under check.

April

Pruning and burning of hanging malformed panicles and proper care of grafted plants should be done. Irrigation of fruit bearing orchards is necessary and spraying of 0.1% carbendazim to control anthracnose in young plants may be done. The population of scales and mealy bugs increases very fast during summer season. Prune out the affected parts and apply insecticides as suggested. Preparation of fruit fly trap (MAT) may be done and it may be installed as suggested. Preparation for harvesting may be done where season is likely to start.

The orchards nearer to brick kilns are especially prone to black-tip disorder. Others may also be affected. Spraying of Borax (1 percent 2-3 times) is advised. Since Borax is not dissolved in ordinary water, it should first be dissolved in a small quantity of warm water. Pruning of affected branches and their prompt destruction prevents the spread of sooty mould. Spraying of mixture of Wettable Sulphur + Gum Acacia (0.2, and 0.3 percent, respectively) keeps sooty mould under control. The Indian oil formulation (tree spray oil, 2 percent) is also effective to control it. Spraying of 2 per cent starch is also effective. During this month, bacterial canker also affects mango fruits. Spraying of Streptocycline @200 ppm keeps it under control.

May

Removal of fallen fruits to control fruit fly, stone weevil and fruit borer may be done. To control fruitflies spot spray of bait (BAT) may be done at weekly interval along with installation of MAT blocks as suggested. If harvesting is likely to start, train the workers and make arrangement of packing materials and necessary items. The population of scales and mealy bugs increases very fast during summer season. Where harvesting has started, proper handling of produce may be done and preservation of fruits may also be done. Prune out the affected parts and apply insecticides in harvested orchards.

Late-maturing mango varieties may be infected by bacterial canker. Spraying of Streptocycline @ 200 ppm (second spray) keeps it under control. Monthly sprays of Bavistin (1000 ppm) or copper oxychloride (3000 ppm) are also effective.

Infection of anthracnose starts on leaves and fruits of late-maturing varieties owing to increase in humidity. The spots due to anthracnose are round, small and necrotic. Spraying of Copper oxychloride 3 g/ litre water during first or second week takes care of the disease.

June

Harvesting of fruits may be continued and arrangement for its marketing should be done skillfully to avoid post harvest loss. If situation permits, hot water treatment may be done. For ripening, appropriate methods may be used. Every care should be taken to transport the fruits safely. Layout and preparation of pits for new orchards and reservation of plants with reliable nurseries for planting new orchards may be done as rains are likely to commence. Nursery man may start preparation of seed beds in the nursery. Replacement of MAT blocks may be done, if harvest is late. Use BAT invariably under such situations. If possible, early harvesting of mature fruits may be done before rain to avoid fruit fly infestation. Orchard sanitation such as collection and destruction of fruit fly infected fruits is required. Sowing of mango stones in nursery beds and proper care of grafted stocks in nursery is to be done. Proper marketing and storage arrangements for the produce of mango and intercrop should be done.

Infection of anthracnose may continue through this month. Red rust may appear on leaves due to high humidity. The spots due to red rust are pink-red, velvety and irregular in shape. Spraying of copper oxychloride 3g/ litre water takes care of both diseases.

The rot caused by *Sclerotium* and *Rhizoctonia* species may spoil seedlings in nursery. Therefore, sterilizing the soil with formaldehyde and then covering with polythene is quite essential. Remove polythene and keep the soil open so that leftover formaldehyde is evaporated.

Mango stones should also be treated with Thiram or Captain (3 g/ litre water) for 5-10 minutes. By doing this, more seedlings are obtained.

July

Harvesting and marketing of fruits to be continued in late areas. Filling of pits for planting new orchards to be continued. Nurserymen should start defoliation of shoots for grafting and grafting to be started if rains have started (at the end of July). Last year's grafts may be taken proper care for sale purposes. This is the right time to sow mango stones in nursery as by this time the stone start to arrive in many areas. On completion of harvesting, start pruning with proper tools. Cut portion must be pasted with Bordeaux paste. After that collect dead and diseased branches from within tree canopy, make a little piles of these pruned branches to burn (it may contain inoculums of certain pathogen). After pruning, spray of any one insecticide at the concentration of 0.06% dimethoate (2 ml/litre of water) or 0.04% diazinon (2 ml/litre of water) to control scale insects, leaf and twig gall midges, shoot borers and leaf cutting weevils may be done which may be repeated after 15 days interval.

Weeding may be done under tree canopy. Clean tillage tools (cultivator and rotovator are preferable) may be used in between tree rows in order to avoid injuries to trees. Never do deep plowing too close to root zone of mango tree. After fruit harvesting rotten fruit is usually left everywhere in a mango orchard. These discarded fruits may contain egg and larvae of fruit flies or it may have fungal pathogens such as anthracnose, soft rot, brown rot etc. This affected fruits must be buried or taken out of mango orchard; by doing this risks of infestation during the next season may be reduced.

The new vegetative flush when it has just started to grow (its tenderness remain intact) may be sprayed with macro/micro-nutrients (boric acid, zinc sulfate). The dosage of above-mentioned nutrients may be given according to leaf and soil analysis results. On completion of harvesting, application of half the dose of each of N, P and K per plant (10 years old and above) on the basis of soil analysis recommendations be done. Mixing of 40-100 kg composted farm yard manure into the basins of each mango plants is suggested. Never use non-composted FYM. Make a 1 and half feet high ring around the base main trunk in order to avoid direct contact of irrigation water.

Mulching is advisable after composting followed by irrigation. This practice will bring earlier vegetative flushes; such flushes will get enough time to get matured for next season. Sowing of green manuring crops such as sunhemp or dhaincha or rainy season intercrops like green gram, black gram, cowpea, etc. Lifting and transplanting the new seedlings in the nursery beds and gap filling in young orchards may be done.

Second and third sprays of copper oxychloride (0.3 percent) at 15-20 days should be done to protect the crop from anthracnose and red rust on leaves.

If some seedlings are affected by rotting, treat the field with captan (0.3 percent). Good drainage facility should be maintained in nursery. Weed out all grasses etc. to remove the collateral hosts of *Sclerotium* and *Rhizoctonia* species.

August

Sowing of mango stones in nursery and lifting and transplanting of new seedlings in nursery for grafting purposes to be continued. Grafting may be started i.e. stone grafting on new seedlings and veneer grafting and soft wood grafting on one year old seedlings. Side by side lifting and sale of grafts may also be continued. In the Terai belt of the country control of shoot gall psylla may be initiated by 15th of August. Two sprays are required after 15 days interval which may be done as per schedule. During this month in certain pockets serious outbreak of mango leaf webber occurs. To control this pest spray of insecticide as per schedule may be done. In case both the pests i.e. shoot gall psylla and leaf webber are present, insecticides which are common to control both may be used. If rains continued planting of new orchards and gap filling may be continued. Application of paclobutrazole in order to check more vegetative flush may be done and proper weed management practices may be followed.

If anthracnose and red rust are not under control, repeat one more spray of copper oxychloride. Clean the field by ploughing deeply as it reduces inoculum of several pathogens. During this period, biocontrol agents (BCAs) such as *Trichoderma* multiplied in FYM should be applied to the basal soil. BCAs will multiply rapidly in the soil, increase vigour and immunity of the plants, and suppress the inoculum build up of pathogenic organisms in the soil.

September

Ploughing of green manuring crops may be done. Second spray should be done to control shoot gall psylla and leaf webber as suggested in August month. If infestation of shoot borer or leaf cutting weevil is observed spraying as suggested in July may be done. Remaining half dose of N and K may be applied, besides, if orchard soils are deficient in zinc, zinc sulfate may be applied that followed by irrigation water. For inducing dormancy, irrigation water may be stopped. (Though it may vary from one place to another). In the nursery, to control anthracnose spraying of 0.1% carbendazim (1 g fungicide /litre of water) is suggested. The second spray may be repeated after 15 days. Planting of new orchards and gap filling may be continued. Lifting and sale of the grafts to be continued. Grafting to be continued. Ploughing, cleaning and removal of weeds to be done.

The incidence of die-back, *Phoma* blight and gummosis increases by this time. Prune infected and dried branches to control die back in orchard and nursery as well. After pruning apply copper oxychloride paste to the cut ends in nursery, whereas on grown up plants spraying of copper oxychloride (0.3 percent) is recommended.

Spraying of copper oxychloride (3 g/ litre water) controls small light brown spots of irregular to larger irregular spots of *Phoma* blight. Similarly gummosis may be controlled by spraying copper oxychloride (0.3 percent).

October

Spraying against leaf webber, if required, may be done. New insecticides may be used in each spray to avoid possible development of resistance to insecticides. Spraying of 200 ppm napthelene acetic acid (20 ml chemical/ 4.5 litres of water) in the second fortnight may be done to overcome mango malformation malady. Completion of lifting and sale of grafts and of planting of new orchards may be done. Gap filling in the orchards to be completed. Mixing of 40-50 kg well decomposed farm yard manure/ compost in the basins of each mango plants. To control die back disease, cut the affected twigs and burn. Spraying of 0.3% copper oxychloride (3g fungicide/ litre of water) is suggested. The spraying may be repeated after 15 days. Ploughing cleaning and removal of weeds to be continued. In the orchards having small plants, sowing of winter crops may be done. To control phoma blight, spraying of 0.2% benomyl (2 g fungicide/ litre of water) or 0.3% copper oxychloride (3g fungicide/ litre of water) in the second fortnight is suggested. Bordeaux pasting to tree trunk should be done invariably.

Second sprays of copper oxychloride (0.3 percent) should be done to take care of die-back and *Phoma* blight diseases. Other necessary fertilizers should also be applied during October. It helps develop the vigour of the plants. Spraying of NAA 200 ppm up to mid-October is quite necessary to keep malformation under control.

November

Spraying against leaf webber may be done if required. Intercrops may be sown and arrangement of grass, bamboo, ropes etc. for making thatches may be done to protect the young mango grafts against frost. Deep ploughing, cleaning and removal of weeds may be done.

Disease control measures in a mango orchard may begin in the winter, when weak, diseased, or dead twigs, branches etc. are removed and are buried or burned. This simple but important practice reduces the amount of potential fungal or bacterial primary inoculum that will start infections in the spring. The orchard floor should be kept clean by removing infected leaves, twigs, collateral hosts, etc.

Third sprays of copper oxychloride (0.3 percent) should be done to take care of die-back and *Phoma* blight diseases.

December

Spraying against Phoma blight if required. Fastening of 25cm wide, 400 gauge thick alkathene strips around the tree trunk about 25 cm above the ground in the third week to stop the ascent of mealy bug *(Drosicha mangiferae)* nymphs. If polythene is not available, make strips of biodegradable grease containing an insecticide (chlorpyriphos) around the tree trunk about 25 cm above the ground. Dusting of 2% methyl parathion dust or 5% carbaryl dust at the rate of 250gm per tree towards the end of the month to control nymphs. Making of thatches over young plants to protect them from the frost and particular attention is to be given to irrigations of young orchards from frost protection. Proper care of intercrops should be taken. Deblossoming may be done to control malformation. Second pruning may be done. It is necessary to spray fungicides soon after pruning. Fungicides against die back, anthracnose, twig blight etc., are already mentioned above. Take a mature shoot for leaf analysis. This is a best time for sampling because the mature shoot will have maximum accumulation of all nutrients. If any nutrient is missing or its ratio is less than normal, further amendments can be easily made. If required, spray with broad spectrum insecticide may be done against scales.

□□□

19

Reduce Harvest, Handling and Marketing Losses

Majority of the mango farms in the country are below 3 hectares (72%). Mango production today is characterized by many contract growing schemes between producers and processors/exporters. The post harvest losses in mangoes have been estimated in the range of 25-40% from harvesting to consumption stage. If proper methods of harvesting, transportation and storage are adopted, such losses can be minimized. The earliest fruit in the market usually commands premium prices (depending on quality). To this end, it is advantageous to pick fruit soon after it reaches maturity. This also ensures that fruit has the maximum shelf life. Fruit left too long on the tree, until it has already started to soften, will not travel well (bruises easily) and lacks shelf life. Immature harvested fruit shrivels and softens but does not develop flavour, and is either rejected at the market place or receives very low prices. Flowering and fruit-set normally occur over a three to four week period resulting in the fruit on any one tree being at various stages of maturity. Depending on the flowering, the crop can usually be harvested in about three picks (for any one variety). Harvesting usually begins with a light first pick of early fruit, about 10-20 % of the crop. The second pick is done a couple of weeks later which is generally the heaviest, about 40-70% of fruit being mature. A final pick is done about a week later which comes from the late set fruits.

Ensure that Harvesting is Done at Right Stage of the Crop

Marketing immature fruit, damages the future sales potential. Mango harvest criteria can vary with local consumption patterns and distance to the market. Time from flowering combined with fullness of mango cheeks are the commonest criteria. Producers must decide whether to harvest as soon as the market price ensures a reasonable return, or to leave the crop in the field to obtain maximum profit. However, mango crop does not permit longer waiting period as it may drastically shorten the marketable life of the produce and lower the sale price. This balance is a critical factor in determining the growers income from the crop. In practice the total harvest period is very short and the grower has very little time to make the correct decision. Mango takes about 3 months from the time of fruit set to mature and ripen. Some varieties may take 3-1/2 to 4 months. If picked immature, fruits develop white patches or air pockets and show lower amounts of Brix, acid ratio, taste and flavour, whereas over-mature fruits lose their storage life. Such fruits present numerous problems during handling. Fruits are plucked from the tree when they are still hard and green. When a few semi-ripe fruits fall from the tree, it is traditionally considered that the fruits are mature for harvest. It is therefore, desirable to pick the fruits at the correct stage of maturity to facilitate ripening, distant transportation and maximum storage life, and thus to increase their quality and market value. Following characteristics are normally used in deciding crop maturity, both are dependent on the experience and personal opinion of the grower.

Maturity Indices

Traditional method

- Mango fruits need 120 to 140 days after fruit set to mature. When a few semi-ripe fruits fall from tree –traditionally considered maturity.
- Cutting open fruit, if the flesh is starting to go yellow as opposed to white, it is reaching maturity. Butter yellow is a good indicator that fruit is ready and mature.

Visual method

- The shoulders are in the line with the stem-end and the colour of the fruit is olive green.
- The shoulders outgrow the stem-end and the colour is olive green.

- The shoulders outgrow the stem-end and the colour becomes light
- The flesh becomes soft and blush develops.

Dipping Test

- Picking fruits when their specific gravity is 1.01-1.02 gives good fruits. Sinking of fruits in water indicates full maturity.
- Fruits dipping in water have Sp Gr. >1.0 or fruits dipping in 2.5% salt solution but floating in 5% salt solution: Sp Gr1.02- 1.04 are considered to be matured for harvesting.
- Fruits dipping in 5% salt solution have Sp Gr>1.04 and are considered to be over matured.

The experiments on storage studies showed that the fruit harvested at any stage of maturity mentioned above ripened normally, but those at the second stage gave the best taste and flavour. Picking of fruits when their specific gravity is 1.01-1.02 is reported to give satisfactory results in respect of uniform repining and keeping quality. The attainment of full maturity, as indicated by sinking in water, appeared to be very essential for normal ripening of fruits under all conditions.

Dry Matter Test

Dry matter test is the best method of testing the maturity of mango, however, it needs some facilities and skill.When a mango has 14% dry matter, it is considered mature enough to pick and will ripen properly with a good flavour. The procedure set out below shows how dry matter percentage can be calculated simply and accurately. Mango dry matter should be used in conjunction with the other maturity indicators of heat sums, internal flesh colour and external colour and shape. The method used to calculate dry matter percentage is to weigh a sample of mango flesh when it is still full of water; then to dry out all the water in the flesh and weigh it again.

Equipment

Cutting boards x 2, large knife, vegetable graters x 2, fork, microwave or drying oven, coffee mugs (for microwave oven), paper towels, patty papers, marking pen and accurate scales (0.1g accuracy).

Method

- Select 3 - 5 of most mature looking fruit for testing. Select samples of the "average" fruit to be tested (avoid obviously forward and backward fruit).
- Record date, variety and block for future reference.
- Cut off both cheeks, cut in half and keep half from each cheek for sampling. Grate samples using a carrot grater onto a china plate. Mix up grated fruit.
- Using two patty cups (one inside the other), weigh exactly 10 g and record. (Use two to prevent the cups sticking to the oven plate). On the beam balance scales, put the weights into two pattypan cups so that the weight of cups is balanced out, using tweezers to accurately weigh in 10 g.
- Place in a microwave with a cup of water (half-full so it does not boil over). Set microwave to medium or medium-low for 30-40 minutes. The setting and the time vary with each microwave. Set the microwave so that the sample is lightly browned not black. The setting varies with the number of samples. Check that the water does not dry out.
- The sample(s) should be monitored frequently. When they begin to go brown and start to dry out they should be weighed every 2-3 minutes.The sample is dry when it no longer changes weight.
- It is important not to burn the samples while drying in a microwave oven. At the same time, the samples should not be under dried. It is desirable to calibrate the technique to the microwave being used. Drying in hot air oven at 65°C for 24 hours is safe and does not burn the samples even if left for a longer duration. This technique is slow but more reliable. Do samples in duplicate, if the difference is greater than 0.5% repeat the test. Accurate scales (to 0.01g) are needed.

Drying Oven

- Place the sample in a drying oven and leave for 12 - 24 hours at approximately 65°C. Higher temperatures will cause burning.
- Record the dry weight + patty paper.

Calculations

- Take away the weight of the patty paper from the wet weight + patty paper and the dry weight + patty paper.
- Record these as dry and wet weights.
- Divide the dry weight by the wet weight and multiply by 100 to bring it to a percentage.

Formula : Dry weight/Wet weight x 100% = % Dry Matter

Do Proper Planning for Harvesting and Disposal

The objective of harvesting is to get the crop without damage and to send it to the market in the best possible condition. Thus, the objective is to move the crop from field to buyer with the minimum number of handling operations compatible with the quality requirements of the buyer; and to minimize exposure of the crop to stresses such as extremes of temperature, or of compression pressures caused by over-loading.

Although the scale of production, availability of labour and type of produce may vary, certain basic factors must be taken into account in the planning of mango harvest operation. Equipment must be obtained, labour organized, marketable produce identified and provision of collection and removal from the field is made. Each of these tasks must be planned, managed and implemented efficiently if the value of the crop is to be fully realised. To achieve this following things are needed:

Good production planning : It ensures that crop maturity coincides with market demand. This can be achieved by proper care while selecting the variety and proper orchard management.

Continuous communication with buyers : It is necessary to identify exact requirements as harvest time approaches, and to let the buyers know the best time of harvest and expected quality.

*Forward planning :*It is necessary to coordinate equipment, labour and transportation facilities.

Field supervision : The efficiency of the harvesting operation itself depends upon the use of experienced or trained staff, and the adoption of methods which will meet the buyers requirements. For this proper field supervision is needed.

Good management of the harvest operations is usually reflected in the speed at which produce moves from field to market place, packing station or storage centre, provided that it is not at the expense of careful handling and subsequent quality downgrading. To avoid this the following steps should be taken.

Give Proper Training to Workers

Training and supervision of labour are critical to a successful harvesting operation. Constant supervision is necessary to maintain quality and reduce subsequent spoilage of produce. Training is required in both general principles and crop specific techniques relating to maturity selection, detachment method, maintenance of equipment, field hygiene and division of labour. Some of the more important areas are:

- Teams of workers must work systematically through a plot or field, experienced staff removing the crop and others carrying it to collection points. If crops are relatively inaccessible, as with older mango, great care must be taken by pickers climbing in the trees if fruit is to be harvested free of damage. For this, planting densities and pruning techniques should be followed which minimize tree size.
- It is essential that crops are harvested at the correct size and maturity for the market. The workers must be given strict specifications before entering the field and their performance supervised carefully.
- During long harvesting sessions, some individuals develop habits of slapping, pressing and rubbing produce. Others become tired and start to throw or drop produce into field containers. Such practices can cause damage to the produce and should be controlled by checking performance, by restricting the lengths of shifts worked, and by promoting comfortable working conditions.
- Sharp edges on rings and bracelets, and long fingernails, are significant cause of post harvest abrasions and should be removed before harvesting starts.
- Unmarketable produce must not be left in the field to rot and contaminate healthy standing crops. Routine collection of waste is an important part of the harvesting operation, and all workers can contribute to this. Cleaning, sterilization or replacement of picking containers must be carried out regularly to prevent the build-up of infection.

- Each individual should be issued once only with necessary equipment and given clear instruction and training in its maintenance. It should then become that individual's responsibility to keep knives and clippers both clean and sharp, and to keep other equipment such as boxes, poles, nets and bags in a good state of repair. Blunt and dirty knives and clippers are potent sources of bacterial contamination.

Decide Proper Time of the Day for Harvesting

The time of day at which to harvest the mango depends on the availability of transport and other facilities, weather and environmental conditions and human factors, as well as market demands. The factor which assumes greatest importance depends upon the crop and the local situation.

- Most favourable condition for harvesting is early in the morning.In some areas, where markets need night transportation, it may be advisable to avoid harvest during the middle of the day. Produce harvested must be balanced against the possibility of local heavy rains which could have detrimental effects. Packing of wet produce often leads to greater post-harvest spoilage and fully turgid tissue can bruise or split more easily.
- Since harvested produce standing in the field usually starts to deteriorate rapidly unless specialized facilities are available, it is unwise to start harvesting until transportation is assured.
- If the crop is to be transported to a relatively distant market, storage centre, packing station or processing facility, harvesting should be timed to allow for delivery at a convenient time.
- Harvesting can only take place when sufficient labour of the required skill and strength is available. Thus the distance which workers must travel, their domestic, religious and sometimes social arrangements may have to be considered.

Picking Techniques and Materials

Adopt proper picking techniques

- Harvest in the morning when fruit is cool and keep harvested fruit out of the sun.
- Detachment of stem at the time of harvest causes fruit latex to trickle down the skin, which imparts shabby appearance to the

fruit by blackening the skin. Hence picking of fruits with stem slightly above the abscission point is recommended. Usesecateurs. Pick with long stems 10 cm or more and lay fruits on their sides in the crates to reduce stem breakage and contaminationby sap.

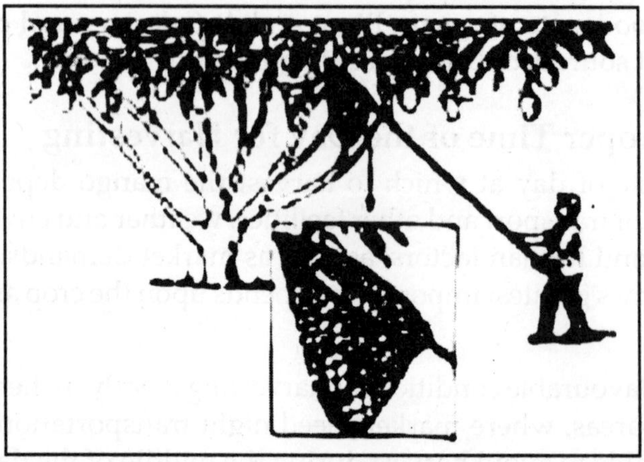

Fig. 22. Proper Method of Harvesting

- Fruits are best harvested using clippers, and placed in harvesting bags carried by the harvester.
- With large trees, fruits are harvested by the use of picking poles, with or without attached clippers, equipped with bags into which the fruit fall. This method is rather slow and requires considerable experience and skill, but is essential if high quality fruit is required.
- Alternatively the fruit is picked by the harvester either on a ladder, or who climbs the trees, and throws the fruit to a skilled 'catcher' on the ground, or into a large net.
- Pulling out of stems from fruit when harvesting has to be avoided at all costs because broken skin at the point of attachment of the stem is particularly susceptible to a decay condition know as stem end rot.
- Injury caused by rough handling contributes to fast ripening, fungus infection and rot. No matter what harvesting method is used, care must be taken to avoid injury to produce.
- Fruit should be handled with care at all times. Hard green fruit dropped from a height greater than 30 cm, onto a hard surface e.g. a branch, will suffer from internal cracking and will not ripe satisfactorily.

- If harvested clean, the produce should be kept clean and not stacked on the soil, even momentarily .Clean the crates and equipments regularly in detergent .
- Have detergent handy to wash hands and equipment in the field and to dip fruit with broken stems in detergents, these fruit should be kept in a separate crate and marked accordingly.
- After rain, postpone harvesting for at least three hours of sunshine to dry out fruit. Skin is susceptible to handling marks after rain.

Use Field Crates and Containers

Harvested produce may be transferred from the harvest containers into field containers, for transport to a packing house, or in some cases for transport direct to local markets.

- Wooden crates made of boards or slats, with a capacity of up to 25 or 50 kg for manual handling. If the wood used in such crates is rough sawn then sharp edges or splinters may cause damage to the contents, especially if the full crates are carelessly handled. Wooden crates are usually returnable or re-usable for local marketing operations. However, rough handling methods usually limit their usage to very few occasions, sometimes as little as three to five return journeys.
- Plastic crates are now in common uses. Most are made from high density polyethylene which contains ultra-violent inhibitor (UVI), which extends the life of the container when it is exposed to sunlight. Plastic returnable crates have a longer life than wooden ones because they are more durable. They are usually made to pack inside each other when empty to save space during transport and storage. Their usefulness as multi-purpose containers makes them liable to heavy losses.
- Sacks and net bags made from natural or synthetic fibres are also used where produce are transported direct from grower to market. They are relatively cheap and require little storage space and usually provide good ventilation to their contents. However, they give little protection to their contents and when made from synthetic fibres tend to disintegrate when exposed to sunlight.

Know the Main Harvesting Season of Mango in the Country

Table 11: Harvesting season of mango in different states

States	Harvesting season
Maharashtra	April-May (Ratnagiri) May-June (other areas)
Gujarat	May-June
Tamil Nadu	April-May
Andhra Pradesh	April-May (coastal districts) May (Rayalseema)
Karnataka	May-July
Bihar, U.P. and other parts of India	June-August
Orissa	April-May (coastal districts) June-July (Hilly districts)

Desapping is Necessary

When mango fruit is plucked , a latex or sap exudes from the stalk close to the base of the fruit. This latex or sap contains mangiferen, resinous acid, mangiferic acid, and the resinol, mangiferol. The latex some times causes blistering on skin and off spots on fruit surface. Intensity of sap burn injuries vary with varieties. Although the damage is not immediately apparent, blemishes usually appear some days later when the fruit arrives at the market.

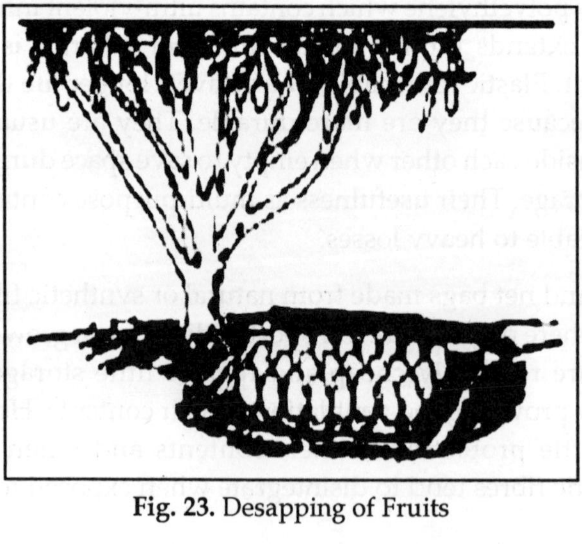

Fig. 23. Desapping of Fruits

There are a number of treatments to reduce the incidence of sapburn.

- Detergent dip is the easiest method. Mango Wash can be used for the purpose@ 200-300 g/100 L of water. The fruit is de- stemmed outside the tank, allowing all the spurt sap to finish before placing the fruit in the tank with the solution. The fruit is left in the tank for one minute (if left in the solution too long skin marking could occur).Because the chemicals in the dip break down during use, it is recommended that 200 kg of fruit will require 100 litres of dip. The dip must then be replaced. Rinse fruit with clean water after removing from the mango wash.
- Fruits harvested with 10 cm pedicels at the time of harvesting and then fruits are placed on the racks keeping stem end down and the stem is snapped off. The fruit is allowed to drain for a couple of hours before packing. So far desapping is done manually.
- The hands of the operator should also be rinsed in detergent to avoid contaminating fruit with sap.

Handling Material from Field to Pack House

Do initial sorting and grading in the field

Much of the required preparation and sorting of produce can be done during the harvest operation itself. Scarred, pitted, over-ripe or otherwise deficient produce can be placed in separate harvest containers or discarded completely and removed from the field later. More rigorous preparation such as washing and critical size or ripeness/colouration selection may be better performed in a pack house . There is a growing tendency towards the practice of packing produce directly into final marketing containers in the field, reducing the number of times the produce is handled with a corresponding reduction in the amount of injury caused. This practice however, requires that appropriate selection and packaging facilities are set up for the purpose, together with adequate supervision.

Collect the produce promptly and safely from the field

Aggregation and bringing harvested produce together for the purpose of further process is necessary. Field assembly should be planned bearing in mind the best location and the provision of basic facilities. Containers filled with produce should not be left exposed to sunlight. As with all aspects of post-harvest, the objective of assembly should be to develop a system which minimises stress on the produce and keeps handling operations to a minimum but with maximum care and to keep the time between harvest and first destination of the produce as brief as possible. Clearly, reaching such an objective does not happen by accident, but only by carefully planning the assembly process as carefully as all other operations are planned.

Transport the produce out of field carefully

To get the produce from big orchard to a collection point by the roadside may involve passing over one or more kilometers of farm roads. Growers rarely give sufficient attention to the logistics of this operation and then wonder why they get problems. In the certain areas this problem is compounded by the topography. In many cases 'heading-out' or mule are the only alternatives and crates may not be as suitable as sacks. If vehicle is used for the purpose it should not be overloaded and it should have good suspension and with preferably larger wheels and lower tyre pressures to take out some of the jolts.

Do Final Grading and Use Ripening Aids

Grading

Once produce is harvested it should be prepared, sorted and assembled together for transport to the market, pack house or storage area as soon as possible. This stage of getting the produce ready for the market should be thoroughly planned, as with all aspects of fresh produce production and marketing.In India the practice of systematic grading of mango, based on scientific grade standards, has not so far been followed by any agency engaged in the distribution of this fruit. The produce is seldom graded in the orchards. At best what the grower does is to remove immature, rotten or diseased fruits from the bulk. Only at the time the wholesalers or pre-harvest contractors purchase the produce, the mangoes are graded according to size and variety before

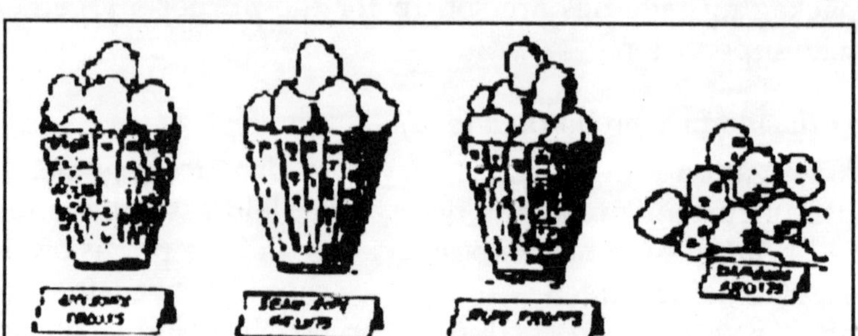

Fig. 24. Graded mango packing

the consignments are booked to the consuming centres. The 'topping' malpractice, where 25% of the better-quality, bigger fruits are placed at the top of each package, is still continued. Grade designations and definitions of quality have been prescribed under voluntary agricultural produce (Grading and Marketing Act, 1937), for 'Alphonso' intended for marketing within the country as well as for export markets, separately. Fruit should be graded and packed into trays according to the following criteria:

- Variety.
- Firm non ripening.
- Uniform size and shape.
- Freedom from blemishes.
- Uniform colour

If the fruits are graded according to their size, weight, colour and maturity, both the producer and consumer are benefited. It has been observed that bigger size fruits take 2-4 days more time in ripening than smaller ones. Hence, packaging of smaller fruits with larger ones should be avoided to achieve uniform ripening. Immature, overripe, damaged and diseased fruits should be discarded. The export quality mangoes are categorised into three grades according to the fruit weight viz., Category-I (200-250 g), Category-II (251-300 g) and Category-III (300-350 g). The fruit can be grouped according to size and quality such as:

Table 12. Class and quality of mango

Class	Quality
Extra Class	Superior quality
Class I	Good quality
Class II	Marketable quality
Out of Grade	Sales of these commodities is illegal

Use Ripening Aids

Mango is a climacteric fruit and its period of ontogeny is characterised by a series of biochemical changes initiated by autocatalytic production of ethylene and increase in respiration. The main changes that occurrs during ripening of fruits is the break down of starch into sugars. There is a continuous decrease in the acidity of fruits during ripening. At ripening TSS content varies from 20 to 24% but after a certain stage, the TSS starts to decline.

It is a general practice to harvest fruits early in the season (premature stage) to capture early market. These fruits do not ripe uniformly without any ripening aid. Use of chemical aids in ripening is mainly intended for faster ripening. Do not use calcium carbide, a banned chemical, for ripening of fruits. Such fruits do not ripe uniformly and quality of fruits is inferior. Calcium carbide is hazardous to health. Ripe fruits with ethylene gas (100 ppm or 0.1 %) in air tight room by exposing them for 24 –48 hrs under controlled conditions of temperature and humidity.

- When using gas, it is necessary to use an air tight ripening room, with temperature control to 20°C and 85% humidity,
- Fruit should be exposed to gas for 24 hours; this triggers the ripening process.
- Only mature fruit should be gassed, the more mature the fruit (dry matter above 14%) the shorter the time to ripening. Immature fruit will change skin colour and get soft, but the flavour will be poor.
- After gassing, fruit must be kept at 18-22°C, even during transport as it may heat up during the ripening process.
- After gassing, fruit will ripen in five to nine days, at these temperatures.

There are a number of methods of gassing fruit:

Shot method

- Ethylene gas is injected into the room at 200 ppm.
- Two shots are applied over the 24-hour period.
- The room must be completely ventilated between shots.

Trickle method

- Ethylene gas is injected constantly into the room at a concentration of 10 ppm.
- It is necessary to calibrate the leakage rate of the room

Alternatively fruits could be ripened uniformly by dipping in 750 ppm Etherel (1.8 ml/litre) in hot water at 52±2°C for 5 minutes within

4-8 days under ambient conditions. Premature fruits (fruits harvested up to 2 weeks prior to maturity) could be ripened to an acceptance quality by dipping the fruits in 750 ppm ethrel solution. Less mature and mature fruits are ripened by dipping the fruits in 500 and 250 ppm ethrel solution, respectively. Biochemical analysis of artificially ripened fruits showed no marked difference with regard to any chemical constituents as compared to the ones ripened by conventional method. The effect of ethrel has been found superior in palatability and with attractive colour development. Such fruits are higher in TSS, sugars and carotenoid content but little lower in ascorbic acid. The ripening process can be triggered in mature fruit causing it to ripen within five to nine days.

This is an option open to growers, which ensures fruit arrives at the market in a ripe condition, and could be useful for air freighted fruit. However, growers must ensure fruit which has been ethylene treated reaches the market within short period, as ripe fruit does not transport well by road.

Hot water treatment (HWT)

Post harvest losses in mangoes in majority of varieties / hybrids have been to the tune of 30%. Mangoes do not normally need any post-harvest treatment for local marketing. Fruits for urban supermarkets may need to be washed if they are heavily contaminated with latex or dust. On a commercial scale, mangoes for export are sometimes dipped in hot water containing fungicide for the control of disease. The treatment is not appropriate for small-scale operations. The major post harvest problems are pests, like fruit fly, anthracnose, fruit rot, and nonuniform ripening. These are causing heavy revenue loss and heavy damage to mango exports. To overcome this problem, IIHR, Bangalore has standardized Hot Water Treatment technology for Mango. "Dipping of fruits in hot water of specific temperature for specified periods for the purpose of disease control, insect damage, disinfestation or uniform ripening is known as Hot Water Treatment". This is also most important in quarantine treatment for exports of fruits. This also has additional benefits like removal of surface pesticide residues, helps to remove sap of fallen in fruit surface during harvest and facilitates washing also.

The hot water treatment plant require the use of a specialized equipment, to control the temperature of the water bath essential for the process. A circulating pump or some other method should be utilized to mix the water to assure uniformity of temperature within the tank.

Temperature monitors should be placed within the treatment tank. The ratio of volume of water to fruiting stock should be large enough to prevent a significant drop in temperature when the stock is added. Sometimes fungicides are added to hot water to bring about efficient post harvest disease control. After treating, the fruit should be allowed to cool rapidly in cold water.

Fruit temperature should not be less than 20°C before starting the treatment. Fruit must be sub-merged at least 4 inches below the water's surface. Water must circulate constantly throughout the treatment. The duration of the hot water dip treatment is determined based on the origin, shape, and weight of the mangoes. It is observed that 46°C for 65min treatment could control both fruit fly and anthracnose and 52°C for 5min. could control anthracnose only.

Packing and Transport

Packing

The most popular containers used for packing mangoes are baskets of various makes, whereas crates, boxes, etc. are used occasionally at some places. Wooden crates are good containers, especially for expensive varieties which are packed in mature condition. The use of steel trunks or tin containers is limited. There is no standardization of weight for any pack and it may differ from place to place. Cost of containers for packing mango also differs from place to place, depending upon the availability of raw material and labour. Padding or cushioning material in packaging of mango varies from grass, paddy or wheat straw, to paper shredding and leaves of mango. In some places, newspapers are used as lining material in different packages. Ventilated wooden boxes with paper shavings in alternate layers as cushioning material have been reported to be ideal for transportation of mangoes long distance by rail .Usually, fruits are placed in layers one above the other, with a straw padding in-between.Wooden boxes are commonly used for packaging and transportation of mango fruits. Central Institute for Subtropical Horticulture, Lucknow has designed and developed CFB Boxes of 5 kg and 10 kg capacity for packing and shipping of mango fruits successfully as an alternative to traditional nailed wooden boxes. The use of CFB boxes for packaging for the domestic market is also the need of the hour due to scarcity of the wood and environmental concerns of the country. For export purposes, CFB boxes are already in extensive use. Polythene (LDPE) lining has also been found beneficial as it maintains humidity, which results in lesser shrinkage during storage. Wrapping of fruits individually (Unipack) with newspaper or tissue paper and packing in honeycomb nets helps in getting optimum ripening with reduced spoilage.

Labeling

Growers packing mangoes for sale on the domestic markets must mark on the end panel of the box the following information:

- The name of the grower or packing establishment together with the address i.e. lot number,
- Street/road and area, not the postal address.
- The mango variety, class and count if packed in mango cartons.
- Date (or date code) on which fruit was packed.
- Fruit in bulk packs must have the weight shown instead of count number.

Transport

With the advancement of transport faculties and road connectivity, now-a-days, mango is transported by tractor or mini trucks even if it within a radius of about 8 km of the market. In almost all the mango growing states, the huge quantities of fruit is transported from distant producing centers mainly by road and rail, a smaller quantity by waterways and only very little by air. In some areas, transport of

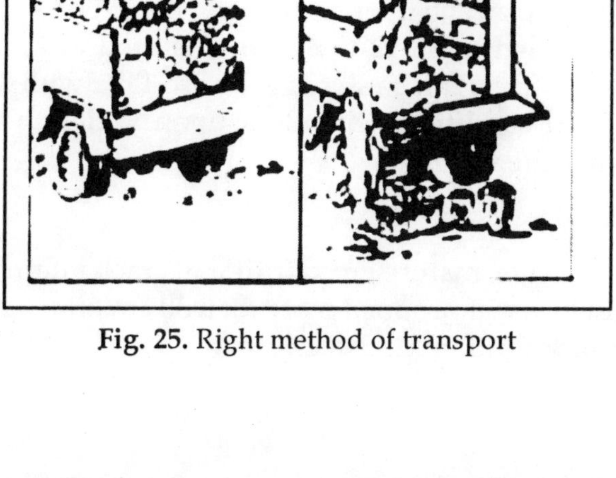

Fig. 25. Right method of transport

mangoes by boats is a common practice. This method of transport is the cheapest. Bombay receives considerable quantities of mango by coastal steamers from Konkan tract from January to the end of May. Owing to distinct advantages in transporting the goods by trucks compared with railways, there has been a definite shift in favour of road transport of mangoes. Trucks are now plying over long distances of 800 to 1,000 km, connecting the producing centres of one state with the consuming centres in the other.

- For local market the harvested fruits are packed in wooden boxes/ CFB and transported by trucks. Packing between vehicles and crates with soft straw or leaves can help cushion the blow but make sure it is clean. Vehicles should be driven slowly. Do not ride on top of the produce.
- Vehicle should be shaded from both sun and rain.
- If the produce is covered with a protective tarpaulin or similar, it should not prevent ventilation of the produce.
- Often the above points are overlooked because the operators argue that the produce does not have far to travel and the produce is not expected to be on the vehicle for long. Unfortunately, drivers do have the habit of making unscheduled stops during their journeys. They may stop for lunch on the way, or to speak to their friends, and often the fact that the pick up is parked in the full sun is forgotten.

Storage and Marketing

Storage

Storage is essential for extending the consumption period of fruits, regulating their supply to the market and also for transportation to long distances. The mature green fruits can be kept at room temperature for about 4-10 days depending upon the variety. For exports, the harvested fruits are pre-cooled to 10-12°C and then stored at an appropriate temperature. The fruits can be stored at 8°C to 14°C with 85-90% for 3-4 weeks in good condition depending upon variety. Other factors affecting storage life are:

- Pest and disease infection
- Impact damage, mature green fruit will crack internally and ripe fruit will bruise if dropped more than 30 cm onto a hard surface. This includes fruit in packages.

- Abrasion damage caused by loosely packed fruit rubbing together during transportation; worn brushes on grading machinery can also damage fruit. The damaged area not only detracts from the appearance of the fruit but can be a source of infection.
- Pressure damage caused by over filling packages, package failure due to over stacking (collapse of bottom layers) and moist cartons losing compression strength when stacked.

Marketing

Mango marketing process is very complicated as being noted in a number and types of middlemen performing various functions of the process.Once mango is ready, its distribution is an extremely important phase in the marketing. After harvest, the fruit has to pass through several channels before reaching the customer. Marketing is a tricky and cumbersome work; hence, generally growers do not go for wholesale distribution of mangoes. They lease out the orchards to pre-harvest contractors-who take care of watch and ward of the crop till maturity and then dispose of the produce as it suits them. Only a small number of producers have direct dealings with the consumers or sell their produce through the commission agents. There is a wide disparity in the prices of standing crop from place to place and even from year to year in the same area and from one orchard to another. It is mainly due to the irregular bearing habit of mango trees, variety, period of maturity and quality of produce. Income from mango growing therefore, is very uncertain which needs to be sustained. Generally commission agents or wholesalers give money to contractors early in the season even before stone formation. Many times growers are in need of money and sell the entire orchard at low rate due to risks. Thus, the farmers are obliged to sell the produce to contractor and contractor is obliged to sell the produce through the commission agents. Sometimes they dispose of the produce directly to wholesalers or retailers. Commission agents, generally known as *arhatiyas* or *dalals* also include the forwarding agents who own the responsibility of proper packaging and transit. They are the most important link in the marketing of mango and about two thirds of the total market is controlled by these agents. They are located in both the assembling (producing) and consuming centers. At some places, they not only sell fruits on commission basis but also transact wholesale business on their own account. Most of the raw material to processing industries is supplied by the commission agents. There are some co-operative societies exist in Gujarat, A.P. Karnataka, Tamilnadu,

Maharastra and Bihar which purchase the farmers' produce and transport it to distant markets to the commission agents.

The retail distribution is done by growers, contractors, commission agents and wholesalers, stall-holders, shop-keepers and hawkers in varying degrees. A fairly large proportion of the profit is taken away by the intermediaries. To ensure better returns to the growers, and fruits at cheaper rates to the consumers, formation of fruit grower's co-operative sale societies deserves encouragement.

Mangoes grown in different parts of the country are transported to the big cities for marketing. The fruits produced in Andhra Pradesh and Tamil Nadu find markets in Nagpur, Mumbai, Kolkata and vice versa. Mangoes for these big markets are usually collected at the central places in all the mango-growing areas, e.g., in Uttar Pradesh, Lucknow and Varanasi; in Gujarat, Gandevi, Gadat and Amalsar and in Maharashtra, Ratnagiri and Vengurla. Delhi and Bombay are the most important markets for dispatch of mangoes.

Price Structure

No minimum support price for mango is fixed therefore; there are no official records or publication showing the prices of mangoes. Further, growers do not maintain any record of their produce and price and the same happens with the contractors. The commission agents and wholesale merchants do keep accounts of their business, but their records provide little information regarding the quality and the variety procured and sold by them. Mango prices vary a great deal from year to year, depending upon each year's total production and various other factors like prevailing prices, demand, transport and marketing facilities. Daily arrivals have a direct impact on the prices. Thus, there is a seasonal and annual variation in prices. Ordinarily, however, the prices are high at the commencement of the season, declining gradually as the supplies increase. Later on, when the arrivals decrease, a high level price is again seen before the close of the season. There is no uniform pattern for price quotations. The unit of sale, both in wholesale and retain trade, varies from place to place. However, to get good price one should make good quality of mango because quality is not a chance but it comes through year round efforts. Further, keeping quality and early or late arrival in the market fetches good price.

❑❑❑

20

Extend Self Life, Reduce Post Harvest Losses and Convert into Value Added Products

Mangoes are climacteric fruits which are harvested at a mature green stage. Post harvest practices are principally concerned and devised to reduce the rate of occurrence, associated with ripening and senescence. Mangoes are the highly perishable. Some times due to over-ripening they have increased susceptibility to develop anthracnose and other bacterial and fungus diseases. Some time internal breakdown, premature ripening of mesocarp in unripe fruit results in un marketable fruits. Similarly presence of insect inside the fruit reduces the reliability and profit of mango growers. Pathogenic decay of fruits and processed products and pest infestation are the important factors affecting the post-harvest quality of mangoes. Post harvest handling and storage technologies needed to control quality of fruit and to avoid post-harvest losses. It aims at to provide to control ripening and avoid or minimize physiological disorders for maximum fruit quality to the consumer. The objective of post harvest management should also include the preservation of mango in the form of various products.

Post Harvest Life of Mangoes

The post-harvest life of mangoes can be divided in to four phases.

1. Storage life (or transport life), which encompasses the period from harvest during which the fruit remains unripe and in condition

resistant to physical damage during normal handling (this phase occupy 4 days)

2. Ripening period, which designates the period from harvest until the fruit attains the stage of maximum consumer acceptability. This period encompasses the storage life period plus the final stages of ripening (this phase occupy 6 days)
3. Shelf life, which start' when fruit is full ripe and is the period in which fruit remain in an edible condition (this phase occupy 6 days)
4. Processed products having long shelf life.

Methods to Increase the Post Harvest Life

Insects, diseases, disorders, nutrition deficiencies, injuries due to transit and storage environment may shorten the post harvest life of mango. All these have to be controlled by proper pre and post harvest treatment and management to give mangoes long shelf life without any loss of nutrients, shape, colour, size and other things. Post-harvest diseases occurs due to high moisture content, vital heat produced due to accelerated rate of respiration, high temperature in transport containers and storage place.

Several pre harvest treatments, cleaning, culling, dipping in various chemicals, antibiotics, oils, hot air treatment, hot water treatment, refrigerated storage and gas storage, fungicides, borax, heated benomyl, oil and wax emulsion have been suggested.

Pre-Harvest Treatment

- Pre-harvest rapping of mangoes results in reduction in fruit fly infestation and anthracnose.
- Pre-harvest foliar sprays of 1% aqueous solution of Zinc sulfate (0.23%) at 20 days interval may increase fruit storage condition for 8-10 days.
- Mango fruits treated with NAA (200ppm) or BA (15ppm) and packing them in polythene bags of 100 gauge with 0.2% vents help in controlling weight loss, maintaining fruit and prolong shelf life by 4-8 days.
- Pre-harvest sprays with chemical fungicides have been shown to reduce post harvest diseases but the effects have not always been

consistent. Benzimidazole, thiabendazole, benomyl, carbendazim, thiophenate methyl are used to control post harvest diseases in mango. The fungicide used to control post harvest diseases should have (a) a high intrinsic activity against many fungi responsible for post harvest diseases. (b) The ability to penetrate, at least, to some degree into the host to reach the site of infection.

- Benomyl at 0.3 g/lt at monthly intervals from flowering to 30 days before harvesting have shown to be very effective. Three pre harvest sprays of 2.0 per cent Calcium chloride at 10 days interval prevents the jelly seed formation (softening of pulp near the stone) and delay the ripening.
- Major post harvest pests are fruit files, stone weevil and mango fruit borer. Fruit fly and stone weevil eggs are laid in the fruits when crop is not even harvested. Therefore, the control of fruit fly and stone weevil may be done in the field itself as discussed in chapter 15.
- Injection of 400g/L formulation of potassium prophonate (15 ml/m canopy diameter) significantly reduces post-harvest stem end rot level in fruit after storage at 22°Cfor 20 days.

Post Harvest Treatment

- There are two safer options for post-harvest disinfestations of mango fruits against fruit fly. (a)After proper harvesting select uniform sized undamaged fruits. Dip them in 5% solution of sodium chloride in cold water for one hour. This will kill 95% eggs in fruit epicarp and also remove the externally present pesticide residues. However, this can't meet international quarantine standard (99.999% freedom).(b)Post-harvest immersion of mango fruits in hot water at 48°C for 45-60 minutes give 100% mortality of fruit fly eggs in the epicarp without affecting fruit quality. The hot water treatment should be done at very small scale initially to develop expertise. A little lapse may cause hardening of fruit.
- Fumigation may be done by ethylene dibromide at 16g / m^3 for 2 hours at temperature of 26°C if fruit fly and stone weevil are problem; however, normally it is not feasible at farmer level.
- Post-harvest decay can be prevented by dipping fruit 24 hours of harvest in hot water alone at 55°C for 5 minutes or for 5 minutes

in hot water not exceeding 52^{0}C (to inactivate deep-seated infections), incorporating a fungicide (1g / 1L benomyle). Use of binomial at 500ul/1 and hot water 52°Cwhen stem-end- rot is only problem. Temperature must be carefully controlled to prevent fruit damage.

- Another alterative is to spray fruits with prochloraz (45-60ml/ litre) applied as a non-recirculated overhead spray for 30 seconds.
- Post harvest treatment of mangoes consisting of 5 minutes hot water dip (50°C) followed by a 20 second ambient temperature prochloraz dip is a conventional method to control post harvest diseases.
- Three minutes exposure to infrared radiation is more effective than minute hot water treatment. Infra Red Radiation is cheaper and faster to control anthracnose, soft brown rot and other diseases.
- Post harvest treatment that involves hot water spray at temperature of 50-60°C but when fruit is applied hot water brush, it significantly reduces decay development of *Alternaria alternata.*
- The rate of ripening in mangoes can be slowed by reducing fruit temperature, but below 10-12°C fruit suffer chilling injury. While the modified atmosphere storage results in to unfavorable fruit responses.
- The Sodium benzoate at 500ppm level inhibits most of the yeasts while potassium sorbet and potassium metabiulphite at 500ppm inhibit all the yeasts. Most of the yeasts affecting mango fail survive he heat temperature at 60°C for 20 minutes.
- The use of ethylene gas -a post harvest treatment to enhance the colour and promote uniform repining of mature mango fruits.

Converting into Value Added Products

Mango processing or preservation is a process in which decay or spoilage of fruits is prevented allowing it to be stored in a fit condition for future use.Mango has many varieties and is grown in majority of the states of the country. Many taste enrichers are made from unripe mangoes and their shelf life is enhanced either by processing them or by using preservatives. Pickles, chutneys and many sweet preparations like murabba are made from unripe or semi-ripe mangoes. This fruit is available only during 4-5 months every year and is generally liked by every one.

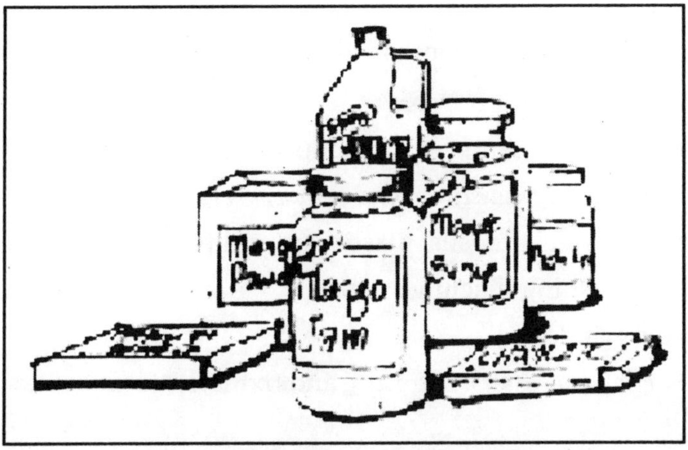

Fig. 26. Preserved materials of mango.

Product and Market Potential

Mango processing is a traditional activity and products like pickle, chutney, murabba are consumed throughout the year. Every region has its own taste or liking and a care has to be taken to understand it and accordingly the recipe has to be finalized. Compliance with FPO and PFA Act is necessary, if processing is done for the market.Indians are fond of table enrichers, which are regularly used along with the main course as well as snacks. Apart from individual households, restaurants, eateries, roadside dhabas, clubs, hostels, caterers etc. are the bulk consumers. There are some branded products available in the market but they are costly. Many Indian households make these items during the season. But this practice is gradually disappearing due to changing lifestyles, hassles of making these items and their availability throughout the year from market. There are many variants of these products and with certain change in the ingredients, taste differs. Hence, it is imperative to cater to the regional palate.

Capital Inputs

Land and building

There is no need to undertake construction by purchasing land. Instead, a readymade premise of around 100 sqm with 2 or 3 rooms would be sufficient for production, storage and packing.

Machinery

Majority of the operations are manual and therefore, production capacity is primarily determined by market. Processing of 6 tones per month for a period of 8 months would require following facilities:

- Gas furnace with burners.
- Cutting and peeling equipments.
- Plastic sealing machine, cap sealer, weighing scales etc.
- Stainless steel utensils, plastic jars and tubs and other tools.
- Bottle washing machine.
- Mixer grinder.
- *Miscellaneous assets:* furniture and fixtures, storage racks, packing tables.

Raw Mango Utilization

Green mango powder

It is also known as mango powder, amchoor powder, dry mango powder, mango powder spice and dried mango powder. Mango powder is a dried spice, extensively used for its taste and flavor in different cuisines. The taste of the spice is somewhat similar to tamarind. It is used for its tangy flavor and a slight tropical aroma. India is one of the largest producers and consumers of dry mango powder. Andhra Pradesh and Uttar Pradesh are one of the major amchoor or mango powder producing Indian states. The dry spice is extensively used in Indian cuisine for giving sour taste to the dish. It is used to make refreshing drinks including jaljeera etc., to make various curries, chutneys and pickles and for tenderizing poultry and meat.Amchur is generally prepared by sun drying slices of mangoes in open place at an

Process flow chart for Amchoor

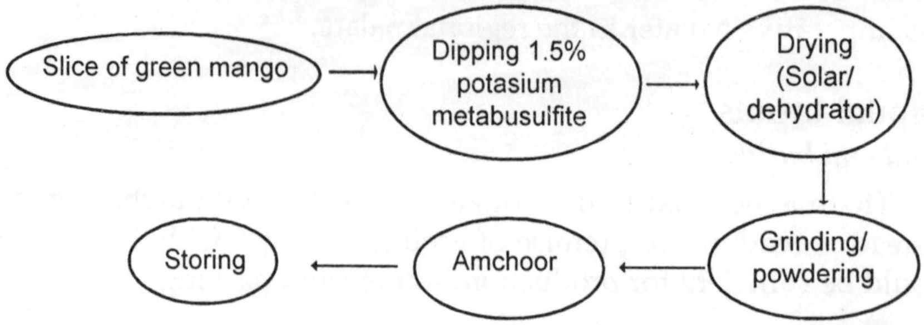

optimum stage of maturity (9-10 weeks after fruit-set). Sulphite treatment of the slices improves retention of colour and vitamin-C.Drying for 15 hr in sun is necessary to reduce moisture content to 2-3%.

Mango Pickle

Pickle is very old method of preserving mango fruits along with salt, oil, spices or vinegar. Mango pickles come in a wide variety of flavors, a mango pickle from South India tastes very differently from that made in North India. In the southern states, sesame (gingerly) oil is preferred, while mustard oil is more typically used in northern India. Andhra pickles and chutneys have a unique flavor and are popular among all those who like it hot. Pickles made from seedling mangoes containing 5-6% acidity are the best in quality. The optimum stage of maturity should be determined for each variety used to make pickles. There are two classifications of pickles - salt pickles and oil pickles. They are processed from whole and sliced fruit with and without stones. Salt is used in most pickles.

Process flow chart for mango pickle

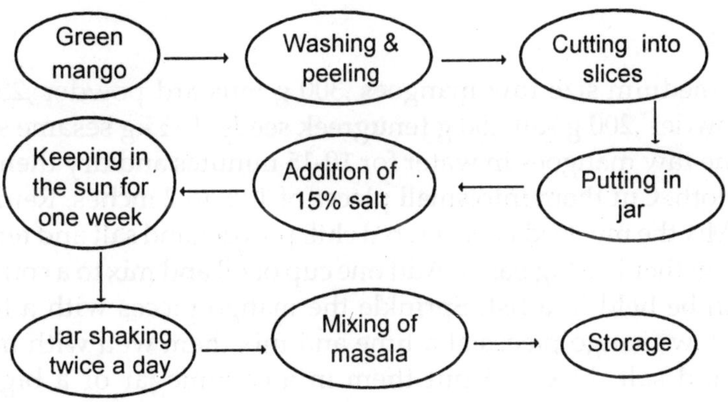

Some of the recipes are given below:

Recipe-1

Medium size raw mangos -8, Mustard oil -4 cups, Mustard seeds-1 tbsp, Aniseed - 1 tbsp., Nigella seeds -2 tbsp., Fenugreek Seeds -2 tbsp, Red chilly powder- 6 tbsp., Turmeric Powder- 2 tbsp., Salt- 1/2 cup. Cut mangoes into small pieces. Put the pieces in a clean, dry earthen jar and add salt to it and shake well. Keep in sun for one day. Mix all the materials - mustard seeds, aniseed, novella seeds and fenugreek seeds, red chilly powder and turmeric powder. Heat mustard oil well until smoky and then let it cool. Mix all the ingredients and salted mango pieces thoroughly by clean dry hands using clean dry spatula. Pour half oil and shake the jar so that all the mango pieces get properly coated with pickle masala and oil. Press firmly and pour the remaining oil. There should be enough oil so that all the mango pieces dip well in oil. Close the mouth of jar with muslin cloth and keep in sun for 12-15 days before use. Stir and shake the jar carefully, at least once in a day.

Recipe-2

One kg raw mangoes,150 g salt, 30 g red chili pepper,3 teaspoons fenugreek seeds,6 teaspoons aniseed, 15 g onions seeds, 7 g turmeric, 2.5 cups mustard oil. Remove the stones and cut the mangoes into big pieces. Rub all the ground spices - salt, red chilly, 12 teaspoons mustard oil and turmeric on mango pieces. Put in a jar and keep for two days in the sun and shake it daily. Then pour rest of the oil and leave for 15 days but not in the sun. Shake it every other day. Serve after 20 days. Be careful to keep mango pieces covered with oil. This pickle is more common in Punjab.

Recipe-3

12 medium size raw mangoes ,300 g mustard powder ,250 g red chili powder ,200 g salt ,50 g fenugreek seeds, 1 ½ kg sesame seed oil. Keep the raw mangoes in water for 10-15 minutes and dry them with a clean cloth. Cut them into small pieces of 1 ½ to 2 inches. Remove the stone. Mix the mustard powder, red chili powder and salt and fenugreek seeds together in a big basin. Add one cup of oil and mix to a consistency that can be held in a fist. Sprinkle the mango pieces with a little oil. Take a few mango pieces at a time and mix them well with mustard, chilly and salt. Mix and put them in a ceramic jar or a big plastic container. Repeat the process till the mix and the mango pieces are over. Towards the end, if a little mix remains it can be added on top. Pour 1 inch oil above it and cover it with a lid. Tie a neat thick cloth over and store it. Mix it well on the third day. Taste it and if need, salt may be added at this stage.

Recipe-4

1 kg raw mangoes and cut them into 4-8 pieces ,1 cup white sea/ rock salt,3 tbsp aniseed/fennel seeds,1 ½ tbsp mustard seeds,3 tbsp

kalonji,1 tbsp fenugreek seeds,5 tbsp red chilly powder,2 tbsp turmeric powder. Put the cut mangoes in the jar and cover with salt. Mix well and cover the jar tightly. Leave the jar out in the sun for a week so that the mangoes soften. Mix the aniseed/fennel, mustard, onion and fenugreek seeds, chilly powder and turmeric together. Add this spice mix to the mangoes. Heat the mustard oil untill it turns smoky and cool it fully. Pour this oil over the mangoes so that all the pieces are submerged. Mix well. Put the tightly covered jar back in the sun for two weeks. Stir everyday to mix well. This method is common in U.P.

Recipe-5

Raw large mangoes - 5, turmeric - 10 g, red chilly powder - 100 g, salt - 1/2 cup, fenugreek powder - 1 tbsp, mustard powder - 1/4 cup, gingerly oil - 2 cups ,fenugreek seeds - 1tbsp. Cut mangoes into pieces and remove seed and pith. Wipe each piece with a cotton cloth so that they are completely dry. In a dry bowl mix all items along with half the oil. Put them in a jar and pour remaining oil; make it air tight with a tight lid. Mix thoroughly on the third day and pour more oil if needed. This method is common in A.P.

Mango Chutney

Mango chutney is a kind of spicy jam which can be prepared from raw mango pulp. It usually contains 55-60^0 Brix TSS and 1.0-1.5% acidity. Mango chutneys are commercially important and they are classified as sweet chutney and hot chutney. Sweet chutney is prepared either from fresh or brined slices. The fruit pieces are mixed with cane sugar and salt and cooked along with coarsely ground spice mixture to a jam consistency after adding vinegar/acetic acid. The hot chutney preparation is similar to that of sweet chutney, except that more spices and less sucrose are used in the recipe. Mango chutney is prepared with the dried/dehydrated slices have better overall organoleptic rating compared to chutney prepared from conventionally brined ones.

Recepie-1

1/2 kg green mangoes, 1 cup vinegar, 1/2 cup water,1 teaspoon red chilly pepper ,400-425 g sugar, 4 teaspoons salt, 2 tbsp long strips of ginger, 2-3 cloves garlic, 3/4 teaspoon big cardamom powder ,60 g Kishmish.Wash the mangoes and dry them. Peel and cut mangoes into very thin and long slices removing the stones. Chop garlic very fine and cut ginger into thin and long slices. Cook the mango slices, garlic and ginger in water on a low fire. When tender, add sugar, vinegar, red chilli pepper, cardamoms, salt and blanched almonds and stir until thick and becomes golden. Remove from the fire, add kiskmish and mix well. Cool and put in an air tight jar. Store in a cool and dark place.

Raw Mango Slices in Brine

Mango in brine is made from sound, clean, raw or sufficiently mature fruit of known variety of mangoes, like Totapuri, Rajapuri. The preparation process includes washing, peeling of semi ripe mangoes, and cutting in slices and stored in 13-16% brine solution with 1 % acetic acid. The process ensures that the natural flavor and aroma of the fruit is retained in the final product. The exotic aroma of the product makes it suitable for making mango chutney and pickle as per own recipe.

Process flow chart for for mango slice

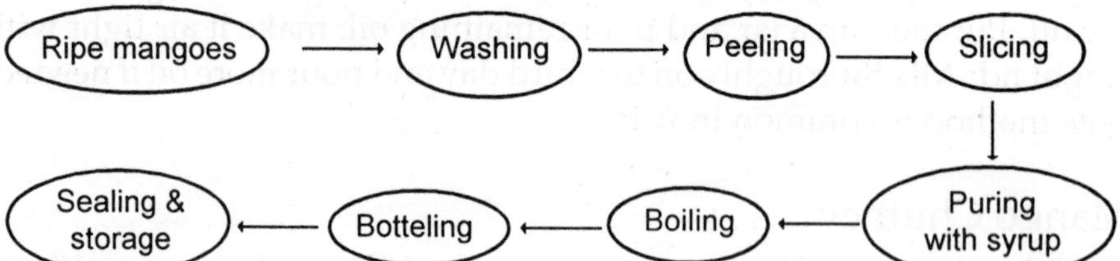

Raw Mango Murabba

Recepie-1

1 tbsp aniseed,750 g sugar,1 tbsp Kishmish,½ tbsp citric acid,½ kg raw mangoes,1 tbsp cardamom powder,8 -9 green cardamom seeds. Peel, cut and deseed the mangoes. Boil in water, when done, throw away the water. Add sugar and citric acid to the water and cook to one-thread consistency. Remove the scum from the surface of the syrup. Add mango slices; simmer till the syrup thickens and the slices become tender and slightly translucent. Leave for 10-12 hours. Remove the mango slices from the syrup. Boil it again till acquires one-thread consistency. Add cardamom powder, kishmish, aniseed and mango slices to the syrup and cook for a few minutes. Remove from the heat and cool.

Green Mango Drink/Panna

It is a popular beverage during the summer in northern India. For preparation of panna, the prescribed ingredients include mango slices, sugar, water, salt, cumin seeds, black pepper and citric acid and mint leaves.

Process flow chart for green mango beverage

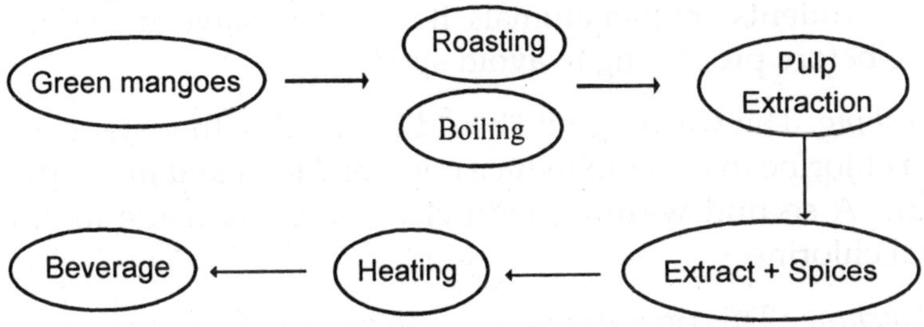

Recipe-1

Six medium sized green mango,3 ¾ th cup water,1 tbsp dry-roasted and grounded cumin seed,1/2 tsp red chilli powder,1 tbsp salt ,3 tbsp sugar,3 tbsp chopped mint leaves,10-12 crushed ice-cubes. Put mango with water to cover in a sauce-pan and bring to boil. Simmer for 10 minutes. Drain off the water and then peel, stone and pulp the mango with a spoon. Place the pulp in a deep bowl. Add water, sugar, salt, chilly and cumin. Whisk thoroughly. Stir in the mint and whisk again. Transfer it to a jug. Serve chilled with crushed ice.

Recipe-2

Green mango slices (grated) -625g, sugar -1 kg, water-1 L., salt-50g, citric acid-10-12 g, green mint extract- 25 ml, black pepper (powdered)- 5 g, cumin seed- 10 g ,potassium metabisulphite- 1 g.Take fully developed large sized green mangoes. Wash, peel, grate and cook with little amount of water. Press and squeeze to extract the juice. Boil black pepper and cumin seeds in 1L of water for 10 min. and strain. Add sugar, salt and citric acid to the strained spicy water. Add mango juice and mint extract along with preservative to the spicy sugar syrup. Mix edible green color (optional) and pack in clean sterilized bottles.

Ripe Mango Products

Ripe mangoes are utilized for preparation of various products like slices, pulp, jam, squash, nectar, juice, RTS beverages, mango leather etc. A key step for preparation of the above products is pulping, as described below.

Fruit selection: Lack of insect infestation, lack of mechanical injuries, proper stage of maturity, uniform colour and texture, minimum soluble solids of 13 ° Brix and pH value of 3.5 to 4.0 are needed for good quality of product. The receiving area must be clean, well ventilated, and free of insects, rodents or other animals. It is not advisable to hold the fruits too long before processing to avoid spoilage.

Washing: The washing pit should be filled with water containing 15 ppm chlorine in order to reduce microbial load and impurities from the fruit. A second washing with clean water is made to eliminate residual chlorine.

Blanching : This operation is done to inactivate enzymes, eliminate air inside the fruit tissues, remove off-flavours and aromas, fix fruit colour and soften the tissues for further pulping. Two methods are currently used to effect blanching: dip in boiling water or direct steam injection. The thermal treatment is applied such that internal fruit temperature reaches 75°C. This usually requires 10 minutes in boiling water, or 6 minutes with steam. Fruit is blanched unpeeled.

Peeling and cutting : Pulp is separated from the seed manually with knives made of stainless steel, on a working bench. Mango pieces are placed in clean plastic containers and taken to the pulping machine.

Pulping : Mesocarp pieces are passed through a fine mesh to remove undesirable particles. After pulping, a smooth puree is obtained. Recommended mesh size is 0.5 mm. coarser material is separated in the process and disposed properly. The pulp is transferred in containers to the kettle.

Additives : The use of additives is recommended to extend the pulp shelf life. Commonly used additives include 0.39 percent citric acid to decrease pH, prevent microbial growth and enhance effectiveness of preservatives as sodium benzoate (0.5 percent). To prevent discoloration, 0.1 percent ascorbic acid is used as antioxidant. Additives are incorporated to the pulp right before the thermal treatment is finished (5 min before) by dispersing in hot water or pulp and proper stirring. Final product should have 13 °Brix and pH values between 3.4 to 3.5.

Packing : The pulp is packed when hot in plastic containers, sealed immediately and flipped over so the internal part of the lid gets in contact with the hot product. All packing materials must be clean before used.

Cooling : Hot containers are cooled with fresh water at the lowest temperature attainable. After cooling, lid closings should be inspected. Finally, containers are cleaned and labels affixed to be sent to a fresh, clean storage place.

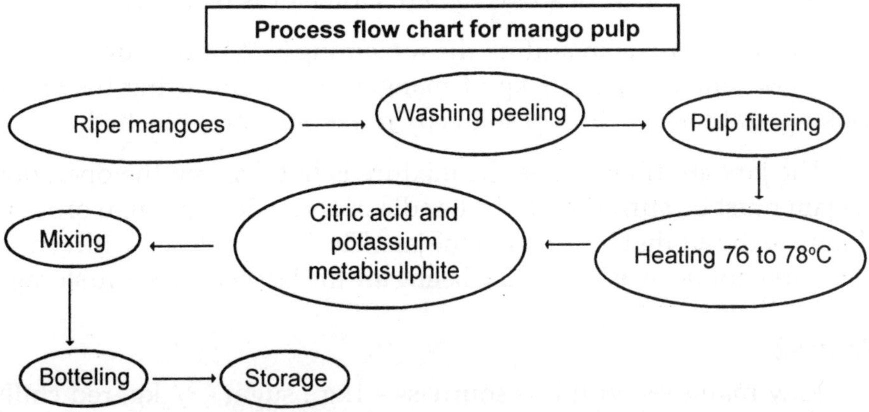

Ripe Mango Powder

It is a high value product and is very difficult to produce a quality product. Freeze dried mango pulp with added sugar increases the shelf life of the powder. Very little loss of flavour occurs, but the cost of production is prohibitive. Foam mat drying results in powder with poor sensory qualities. Spray drying mango pulp produces powder of good colour but lacking in flavour. Incorporation of sugar, milk solids, glycerol monostearate and alginate to mango pulp results in good quality spray dried powder.

Mango Jam

Jam is a processed product in which fruit pulp and sugar is cooked till it is converted into thick syrup. A good quality jam consists 0.5 – 0.6% acidity and 40% sugar. Total soluble solid (TSS) of jam is 68-70%.End point of jam is checked by sheet test or drop method. Microbial spoilage in jam can be checked by adding 40 ppm sodium metabisulphite.

Recipe-1

Both ripe and under ripe fruit are used. The mangos are washed, peeled and cut into small slices with a stainless steel knife. The amount of sugar required represents 60% of the weight of the mangoes prepared. The cooking is done in two stages:

1st step consists in adding 70% of the amount of sugar calculated, plus 2 spoons of lemon juice per kg of mango. Stir well during the entire cooking until 50° Brix of solids by refractometer is reached.

2nd step consists in adding the remaining 30% of the sugar plus 2 spoons of lemon juice per kg of mangos. Stir well during the entire cooking, until 67-68° Brix of solids by refractometer is reached.

The jars are filled while the mixture is hot. During the operation the jam must be stirred with the handle of a wooden spoon in order to eliminate the air that has entered the jars. The jars are closed with screw-tops. Jars cooling, washing and labeling are the last stages before storage.

Recipe-2

Raw mangoes with less sourness - 1kg , sugar - $^3/_4$kg, red chilly powder-$1^1/_2$tbsp, fennel - 1tbsp, cumin - tbsp,water - $2^1/_2$cups. Peel the skin of the mangoes and cut it into small long pieces. Mix the sugar and the water. Keep it on fire and stir continuously. When it boils, add the mango pieces into it. There must be sufficient amount of water to immerse the mango pieces completely. Close the vessel with a lid, simmer the fire and cook for about 10 minutes. When the mango pieces are done, smash it with a fork. Fry the fennel and grind it. Add the fennel powder and chilly powder into the mango solution. When the solution thickens, add the cumin seeds. Stir well and remove from fire. When cool, pour it into a bottle.

Recipe-3

Mangoes (getting ripe)-1kg, clove-12, cassia (1 inch long) or cinnamon - 6, sugar-$1^1/_2$kg, citric acid - $1^1/_2$tbsp, yellow colour- 2 pinch, potassium meta bisulphate- 2 pinch. Peel the skin of the mangoes. Cut into small pieces and put in water. There must be enough water to cook the mango. Slightly crush the clove and cassia and put this into the water. Heat it. When it boils well, remove the cassia and clove from it and switch off the fire. Grind the mango in an electric liquidator, when it cools. Replace the mango solution in the fire. Add sugar and citric acid to it. When the jam becomes thick, add the yellow colour dissolved in 1tsp of water and remove from fire. When the jam becomes cool, dissolve potassium metabisulphate in half of its quantity. Mix it with the other half and store in a bottle.

Recipe-4

Mango pieces - 500gms, sugar - 250gms, lime juice- 2tsp,water - 1cup. Remove the skin of the mango and cut it into small pieces. Cook it with the water and smash it well. Add sugar and lime juice. When it becomes a thick paste, remove from fire and pour the content into a bottle. To test whether the jam is done, cool the jam on a ladle. When cool, the jam should fall from the tilted ladle in flakes rather than in drops.

Mango Slices

Slices (peeled) obtained from ripe and semi-ripe mango fruits are preserved in sugar syrup. Varieties with fine texture and strong aroma are preferred for making the slices. Half-ripe fruits, without fibers, are used. The mangos are washed, peeled and cut into 6-8 mm thick slices with a stainless steel knife. To obtain finished products with good quality and long storage life, the mango slices are soaked for 18 hours in a solution containing boiling water-1 litre, sugar up to 800 g, potassium metabisulphite -3 g/litre of water and lemon juice- 2 spoons/litre of water. The slices thus prepared are drained and placed on glycerin coated aluminum trays, which are placed in a sun or solar dryer. The drying is completed when the product has a moisture content of 15%. The dried slices (150 g) are packed in cellophane bags, labeled and stored in a dry place.

Mango Juice

Fully ripe fruit is used. The mangos are washed, peeled and cut into slices with a stainless steel knife. Pulp extraction is carried out with a hand-driven or electrical juice extractor.

Boiling water, lemon juice and sugar are added to the pulp so that the mixture contains 12% TSS as determined by a refractometer and pH of 3.5 to 3.8. The composition of ingredients is as boiling water- 1 litre/kg of pulp, sugar- 200 g/kg of pulp and lemon juice- 2 spoons/kg of pulp. Bottles are filled and capped with a manual capper. A pasteurization time is related to the size of bottles. Allow the bottles to cool in the same container till the following morning then wash, label and store them.

Mangoes product such as nectar, juice, pulp, slices are packed in container -tin free sheet and flexible plastic pouches during storage. The plastic bags are vacuum closed and frozen in cold room at 20C$^{\circ}$ up to 120 days.

Process flow chart for mango juice

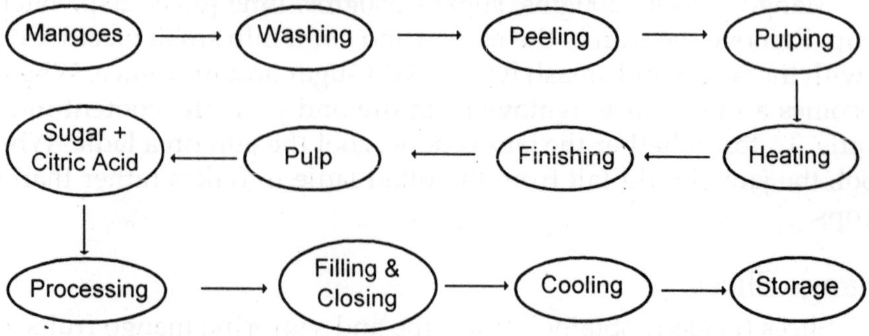

Mango Leather and Bar

Mango leather or 'Am Papad' is another popular product which is prepared by drying mango pulp to around 15% moisture. Ripe fruit is used for the purpose.The mangos are washed, peeled and cut into pieces with a stainless steel knife. Pulp extraction is carried out with a hand-driven or electrical juice extractor. Sugar, lemon juice (or citric acid) and potassium metabisulphite are then added to the pulp so that mixture contains 25% TSS as determined by a refractometer. The composition of ingredients is as sugar- 10 to 15% of the weight of the pulp, lemon juice- 2 spoons per kg of pulp and potassium metabisulphite 2 g per Kg of pulp. The pulp thus prepared is heated for two minutes at 70-80° C. It is then poured into aluminum trays coated with glycerin (this facilitates the removal of the dried pulp). The prepared pulp should be placed on trays at the rate of 15 kg per square metre. The trays are brought to a sun or solar dryer. The drying is completed when the product has the consistency of leather (about 15% moisture content). The yield ratio between raw material and finished product is about 12:1. Two or three layers of dried product are piled one on top of the other and cut into small squares (4 X 4 cm). Each square is wrapped in cellophane paper, packed in cellophane bags then labeled and stored in a dry place. The ideal moisture to have storage stability is 15% or a little more with a relative humidity between 63-70%.

Mango Cereal Flakes

Mango pulp is blended with small amount of wheat-flour and sugar at a fixed pH. The blended and homogenized mass is dried on an atmospheric double drum drier; rotation of the drum is adjusted to 2-3 rpm with the clearance between the driers of 0.35 mm. The product, which dries in the form of a crisp continuous sheet, is broken into smaller flakes and packed into air-tight tins.

Mango Squash

Squash is prepared by mixing 1 kg of pulp with sugar syrup (1 kg sugar in 750 ml water). The whole mixture is heated to 76-78°C and then 25-30 g citric acid is added to it. The prepared squash is filled in clean, sterilized bottles and stored. At the time of use, three parts of water is added to one part of squash.

□□□

21

Identify the Problems and Select Appropriate Remedy as and When Needed

This section of the book has been specially included in order to enable the farmers who are not able to take appropriate action against a problem owing to confusion. Quite often, farmers encounter these problems and due to lack of proper guidance, they fail to rectify the situation even after spending the money. The diagnosis given hereunder is indicative, for more detail information can be had from the respective book chapters.

Problem: Young Plants Die Suddenly

- If sap oozing is seen and bark is dead cut the affected portion of plant. If big size creamish white larvae with dark brown colour head are present, there is every chance that plants are dying due to attack of stem borer.
- If new orchard is located in cold areas and plants are dying in winter, it may be due to frost injury.
- Dig out the dead plant and see if the roots are coiled. If so, death of young plant may be due to goose neck of planted sapling.
- If no symptoms of borer and frost injury are seen, there is a chance that plants are dying due to a complex problem of tree decline or *verticillium* wilt.

- Termite may attack on root and destroy it, resulting in plant death.
- If orchard is located in dry areas under light soil condition with no irrigation support, there may be a chance of wilting due to severe water stress.

Problem: A Good Fruit Set was Obtained but it is Falling

- It may be the natural phenomenon to keep balance between source and sink or it may due to hormonal imbalanece.
- There may be moisture stress, especially in light soil and dry wind prevailing area. This may lead to heavy fruit fall.
- Examine the dropped fruits properly. If hole is present on nose of fruit, there may be attack of fruit borer. If not, cut open the fruit and see wheather there is presence of white larva in stone? if yes, there is attack of stone weevil. Due to infestation of these two insects, heavy fruit fall may occur.
- Heavy attack of sap feeding insect such as, hopper and mealy bugs also leads to heavy fruit fall.
- Cut open the fallen fruits. If no seed is seen, there is possibility of poor pollination and fruits are not developing further and falling.

Problem: Young Plant Remains Unchanged/ Not Attaining Growth

- Soil is low in nutrient and poor fertilizer management have been adopted.
- Dig out the plant and see if the roots are coiled/goose necked. If so, plant may not attain growth due to poor supply of nutrients and water to aerial portion.
- There may be hard pan below the soil and proper pit was not prepared while transplanting the sapling or the soil may be salt affected.
- In certain cases, rootstocks do not support the scion to develop properly due to incompatibility and growth is restricted.
- Small pit was made in problematic soil while planting the sapling.

Problem: Plant age is More than 5 Years but not Flowering

- Soil is extremely rich and enough nutrients are already available therefore, additional fertilizer dose is triggering vegetative growth.
- If water table is high or soil moisture is too high due to irrigation especially during flower bud differentiation stage, vegetative flush may appear in place of flowers.
- Whether such intercrops are grown which require irrigation during flower bud differentiation?
- The planted sapling in not grafted. It may be a seedling plant which take more time to flower.
- The soil is very poor in nutrition and no management practices are followed.

Problem: Poor/No Fruit Set and Heavy Flower Drop

- Too low temperature may lead to poor pollination and thereby low fruit setting. Mango is highly cross pollinated crop; hence, poor activity of pollinators also leads to poor fruit set.
- Most of the flowers are male flower and as matter of fact male flower will drop.
- The major reason of flower drop is attack of insect pests such as flower midge/ anthracnose/powdery mildew/mealy bug/ hopper.

Problem: Poor Germination of Stone in Nursery

- The stone sown in the nursery may be diseases affected. Such seeds have black kernels.
- The stone are affected with stone weevil and kernels are already eaten up by the insect.
- Old stone stock loose their viability after certain period of time. Sowing of such stones leads to poor germination.
- After pulping the stone was left in the sun and kernels are dried. The dry kernel loose viability.
- High moisture in seed bed may lead to rotting of seeds especially where poor aeration is prevailing.
- The seed bed is termite prone and stone is attacked by termites.

Problem: Poor Growth of Rootstock

- Poor nutrition in potting mixture may not support good growth of root stock.
- Heavy leaf fall due to leaf cutting weevil/anthracnose may reduce the growth of root stock.

Problem: Scion Dies but Root Stock Remain Alive when Planted

- The graft union was not matured enough to be fit for field planting.
- The graft was taken out for planting without hardening.
- High temperature and bright sun at the time of planting.
- Soil moisture not enough to compensate the water loss by graft.

Problem: A Portion of Plant/Branch Dies Suddenly

- Stem borer attack may lead to death of part of plant.
- Tree decline/disease may cause the death of a particular part of the plant.

Problem: Fruits Attain the Size of Lemon Get Yellow and Fall

- Too low temperature may lead to poor pollination and thereby low fruit setting. Mango is highly cross pollinated crop, hence, poor activity of pollinators also may lead to poor fruit set. Some of the fruit fall off early where as others develop to egg-size but get yellow and fall subsequently. These fruit are seedless commonly called nubbins.
- Mango flowers set many more fruit than the tree can hold many of them get yellow and fall due to shortage of food materials.

Problem: Flowers Become Black and Fall

- The flower is heavily attacked by midge/hopper/mealy bug. The secretion of these insect leads to blackening of flowers.
- The fungal disease anthracnose/powdery mildew or bacterial black spot may blacken the flower.

Problem: Young Leaves Going Brown in Rainy Season

- The attack of anthracnose causes leaf browning.
- Excessive dose of fertilizer may lead to leaf browning.
- Saline water irrigation may cause leaf browning.
- Leaf miner make tunnels and turn leaf to become brown and dry.
- Toxicity of some nutrients amy cause leaf browning.

Problem: New Growth is Wilting

- The insects suck sap from new growth, which causes grooves along the length of the new stems, causing them to wilt. The mango shoot borer hollows out the inside of new stems which leads to wilting of new growth.

Problem: Fruit is Cracking

- Attack of bacterial black spot or anthracnose may cause fruit cracking.
- Deficiency of boron leads to fruit cracking.
- Severe water stress and sudden irrigation disturbs the water balance which leads to fruit cracking.

Problem: Mangoes Remains Green After Ripening.

- Many varieties retain green color skin even after full maturity.
- Incorrect ripening practices may lead to green skin colour.
- High nitrogen levels in the tree may retain green colour in fruits.

Problem: Fruit Skin Becomes Brown

- Sap coming out of stalk at the time of plucking of fruits cause burning and browning of skin.
- Poor aeration, excessive moisture supplemented with heat causes skin browning.

Problem: Regular Bearer Plant but some Years don't Flower.

- Mango trees are sensitive to climatic conditions at flower initiation.
- Unavailability of dormant period in winter to initiate flowering.

- Reduced carbohydrate reserves in the tree due to a very heavy crop in the previous season.
- Late pruning causes late emergence of shoot which do not mature for flowering in the forthcoming season.
- Irrigation/ rains at flower bud differentiation may cause vegetative growth in place of flowers.

Problem: Orchard is Established in Low Depth Soil/Poor Soil

- Keep plant dwarf and use vacant spaces for intercropping
- Go for high density planting by planting more saplings in between the rows.
- Adopt moisture conservation practices such as mulching.
- Ensure proper irrigation and nutrient for plants.
- Enrich the soil with green manuring.
- Ensure proper irrigation for plants and best possible nutrition.
- Apply fertilizer in split dose and take special care of micro nutrient management.

Problem: Orchard is Grown in High Water Table and Flood Prone Area

- High water table and flood cause many problems in mango, hence, make appropriate provision of drainage.
- Mound the soil around the trunk. This will permit to grow feeder roots above the water logged soil surface to respire under submergence situations.

Problem: Orchard is Under Harsh Temperature Regime

- Provide mulch to check the evaporation and root baking in summer.
- Establish thick wind break.
- Provide irrigation at critical stages.
- Thatching of plants in early stages to prevent hot/cool air.

Problem: Orchards is Located in High Wind and Cyclone Prone Areas

- Establish thick wind break to check the high current of wing blow.
- Dwarf varieties should have been selected, however, if not, keep the plant stature dwarf by pruning.
- Make deep pit at the time of planting to allow deep penetration of roots.

Problem: Orchard is Established with a Wrong Variety / Seedling Plants have Emerged out of Grafted Plant/Mix Scion Block was Developed

- Replace the plants, if few are of undesired variety.
- Head back the plants and go top working as the process adopted in rejuvenation of old orchard to get canopy converted into desired variety.
- Plant may be removed and stones may be sown for *in situ* grafting.
- The individual plant should be properly labelled and collect scion carefully if it is used as mother plant to avoid mixture of saplings.

Problem: Rootstock Dies when Transferred in Pots or Secondary Nursery

- Do not delay in rootstock shifting. Early shift gives good rate of survival of rootstock.
- Prevent tape root while shifting of saplings.
- Do not over irrigate the medium and make it favourable for rootstock
- Keep straight roots while planting
- Follow proper care of shifted rootstock.

Problem: Scion Dies After Grafting

- Proper grafting method is not adopted. Give training to grafter.
- The thickness of scion is not matching with rootstock, thus, cambium layer is mismatching.

- Seasoning of scion has not been done before grafting.
- Immature scion has been selected.
- Grafter is removing all the leaves from rootstock which leads to failure of graft.
- Grafting is being done under unfavorable condition such as too high or too low temperature and extremely dry condition.

Problem: Grafts do not Attain Growth in the Nursery.

- Attack of leaf cutting weevil or anthracnose.
- Poor potting mixture.

Problem: Leaves Burning /Turning Black

- Attack of anthracnose may cause burning of leaves.
- Chloride toxicity may burn the leaves.
- Heavy attack of flower midge, scale insect, mealy bugs of hopers may lead to development of shooty mould on leaves.

Problem: Seedling Become Goose Neck/Root Coiled

- Potting bag is too small which is not permitting roots to grow.
- Saplings have been kept for longer period in potting bags.
- Proper care was not taken while shifting the rootstock and tape root was placed wrongly in potting bag.

Problem: Fire is Matter of Common Occurrence

- Have a 5-10 m wide fire belt around the orchard plowed and harrowed to remove dry grass.
- During the dry months of the year keep an area of about 3 m around the base of trees weed free.

There are many more such queries, through which, a farmer may come across. Such problems may be analyzed and help of the experts may be taken.

□□□

22

Appendix

1. *Trichoderma* Enrichment of Farm Yard Manure

Trichoderma formulation can be used for enriching farm yard manure (FYM) by applying 2 kg of the any one of formulations to one ton of FYM. The FYM should be well rotten and having appropriate moisture, neither too moist nor dry. The FYM should be spreaded on the ground and *Trichoderm*a powder should be mixed properly by turning with spade. Then a heap should be made and covered with paddy straw. This should be left for 15 days under the shade. It is advisable to maintain optimum moisture for the better enrichment of FYM with the bio-agent. After 15 days the FYM can be used for applying to the orchard. It helps plants in many ways.

2. Method of Leaf Sampling for Nutritional Analysis

The aim of leaf and soil analyses is to determine the nutrient status of mango trees or suitability of a soil for the production of mangoes. On the basis of leaf and soil report, one should go for application of nutrients.

A single leaf or soil sample should be representative of an area not greater than 3 ha. However, if there are soil variations, separate leaf and soil samples must be taken and the orchard management adapted accordingly.

How to take representative leaf samples

- The time of leaf sampling as well as leaf position, is very important. Leaf analysis is only applicable for producing mango trees (normally a tree age of 5 years and older).
- Select about 20 healthy trees by walking diagonally from the corners through the orchard. The trees should be homogeneous in appearance and representative of the orchard. Exceptionally good or poor trees must not be sampled.
- The 20 selected trees must be clearly marked, for example, with paint, so that both the soil and leaf samples can be taken from the same trees every year. Sample should be collected from 7 months old fully developed hardened off leaves from fruit bearing twigs
- Where possible, pick 4 leaves from alternate sides of the tree at about shoulder height. Eighty leaves per sample should be sufficient. Different cultivars should be sampled separately. Leaves sampled must be free of sunburn, disease and insect damage.
- Leaf samples should be collected in the morning, after the dew has dried off. Leaf samples should not be taken if trees are under stress i.e. drought or high temperatures. After a heavy downpour, wait at least 2 weeks before taking samples. After sampling, leaves should be placed in clean, perforated or open plastic bags.
- If samples cannot be delivered immediately (within 48 hours), they can be stored in a refrigerator and should be transported in a cooler bag. The sample must be accompanied by the relevant orchard information including previous production figures, tree age and fertiliser programmes of the past. Any problems oncerning the specific orchard, such as small fruit, should be mentioned.

The critical standards for leaf nutrients have been generated for mango plants. Based on the analysis report and standard, the framers should decide the dose of nutrients.

Critical leaf nutrient standards

Nutrient	Standard
N (%)	1.23
P (%)	0.06
K (%)	0.54
Ca (%)	1.71
Mg (%)	0.91
S (%)	0.12
Fe (m g/g)	171.0
MN (m g/g)	66.0
Zn (m g/g)	25.0
Cu (m g/g)	12.0

Source : Bhargava and Chadha (1988)

3. Method of Soil Sampling for Nutritional Analysis

The sampling depth: Top soil 0 - 300 mm, Sub soil 300 - 600 mm

Number of samples : A sample comprises of a combination of at least 10 subsamples. A composite sample should not represent more than 3 ha. Samples from different orchards or lands should not be combined.

The sampling points : Take samples by walking diagonally from the corner through the orchard or land. In an established orchard, topsoil and subsoil samples should be taken at the same trees selected for leaf sampling. Soil samples must be taken under the canopy of trees in the middle between the stem and the drip area perimeter.

The sampling method: Clear the soil surface of debris, leaves and fertiliser. A soil sample must not be taken too soon after fertilising because this will contaminate the soil sample and lead to an incorrect analysis. The top and subsoil samples are taken by removing a core of soil from the top 0 to 300 mm and then from 300 to 600 mm soil depth, respectively.

Packaging of samples : Sub samples from an orchard or land should be combined in the respective bucket (not a fertiliser bag) and mixed thoroughly. A sample of about 2 kg is taken from the composite sample and dispatched in a clean, strong bag. Topsoil and subsoil must be packed separately and the depth of sampling, orchard/land must be indicated clearly.

4. Preparation and Use of Bordeaux Mixture

The Bordeaux mixture is very effective in controlling several diseases of mango. The growers can themselves prepare and use the Bordeaux mixture in their orchards.The Bordeaux mixture can be easily prepared by mixing copper sulphate solution and quick lime suspension in a fixed ratio (2:2:250) i.e. copper sulphate (2kg), quick (unslaked) lime (2 kg) and water (250 litres). The raw materials should be of good quality. Copper sulphate (98 per cent purity) and the freshly stocked lime stone (quick lime) should be obtained.For preparing the 2:2:250 Bordeaux mixture, dissolve 2 kg of copper sulphate in 125 litres of water. Copper sulphate dissolves slowly in cold water. Hence, in winter it is better to use hot water or suspend copper sulphate in a muslin cloth bag into a non-metallic vessel (wooden barrel or earthen pitcher) so that the material gets dissolved in water and solid impurities, if any, are left out. Take 2 kg quick lime in another vessel (earthen or metallic). Avoid plastic container for this purpose. Add some water to the lime to make a paste to which water should be poured slowly to make the volume to 125 litres. Stir vigorously to make a lime suspension.Mix the copper sulphate solution and lime suspension by pouring them together in the third vessel constantly stirring the mixture with a wooden stick while pouring. The lime solution should be poured through a strainer to keep back all lumps. A light blue liquid is obtained which should be strained through a muslin cloth or sack before putting in spray tank.

Uses of Bordeaux mixture

The application of the Bordeaux mixture has been recommended for the control of a range of fungal and bacterial diseases such as leaf spots, blights, anthracnose, downy mildew and cankers etc. It, however, causes burning of foliage or russetting of fruits such as apple when applied under extreme weather conditions. The phytotoxicity of the Bordeaux mixture is reduced by increasing the ratio of lime.The Bordeaux mixture also acts as pit disinfectant in root/collar rot infested sites before planting new saplings in orchards. When used as a basin drench treatment, it controls the spread of root rot pathogens from diseased plants to healthy ones. It renders protection to pruning cuts or injuries caused to the spurs during harvesting when used as a prophylactic treatment.

Test Bordeaux mixture before use

Performing any of the following tests can test the safety of the Bordeaux mixture:

Litmus test: The Bordeaux mixture is alkaline in reaction as it contains more proportion of lime. It should, therefore, turn red litmus paper blue. An excess of copper compound in the mixture may be dangerous to foliage of many plants and is indicated by solution turning blue litmus paper red.

Copper deposition test : It is a simple and practical way of estimating whether or not the mixture contains excess of copper in it. Immerse a bright iron surface such as knife blade or a nail in the upper layer of the liquid for a few minutes. A brick red or bright pink coppery deposit on the metal surface indicates the presence of excess copper and as such the solution is unsafe for use. Therefore, more lime solution should be added to the mixture and the test repeated till no such deposit appears on the clean metallic surface.

Reagent test : A more precise chemical test may be performed using the reagent, potassium ferrocyanide. The reagent may be obtained from a scientific chemical store. A few drops of the Bordeaux mixture are added to little of this reagent in a saucer. If no change occurs, the mixture is safe for use but its turning reddish brown, indicates the presence of excess free copper in it, which is not desired.

Some do's and don'ts in use of Bordeaux mixture

- Use the Bordeaux mixture soon after preparation. If should not be stored for further use.
- Do not use metallic containers for preparing copper sulphate solution and lime suspension. Use a wooden stick for stirring to get a homogenous mixture. Never use metallic stick.
- Do not use the Bordeaux mixture in combination with any other chemical or pesticide keeping a gap of at least one week from that of the Bordeaux mixture.
- To avoid choking of the nozzle, it is advisable to strain the Bordeaux mixture through a cloth or a sieve before putting it into the spray tank.
- The Bordeaux mixture tends to sediment easily. Therefore, its stirring while using is desirable.
- In exceptionally hot days, when the plants are showing signs of temporary wilting or when it is raining, the Bordeaux mixture should not be sprayed, particularly on newly emerged tender foliage.Exercise this caution particularly on nursery plants.

- After carrying out spray operations the appliances should be thoroughly washed with plenty of water to remove any copper deposits.
- The left over Bordeaux mixture should not be dumped in the field.

5. Bordeaux Paste

Pruning is a regular horticultural operation used of removing dead wood, diseased parts and to give shape to the fruit tree. It is undertaken annually during dormant period. The cut ends created during pruning need to be protected against infection by different pathogens, by sealing with suitable fungicide pastes. The Bordeaux paste is the most effective wound dressing material for this purpose. The fine layer formed on the cut ends kills the pathogen, if any, protecting it from subsequent invasion and results in rapid healing of the wounds. Application of the Bordeaux paste can effectively check the attack of collar rots and root rots of fruit trees.

Preparation of Bordeaux paste

The ingredients for the Bordeaux paste are copper sulphate (2 kg), quick lime (3 kg) and water (30 litres).

Dissolve 2 kg of copper sulphate in 15 litres of water. Slowly slake 3 kg of quick lime in a small amount of water. When slaked, add enough water to make the volume up to 15 litres. Thoroughly mix these concentrated solutions of copper sulphate and lime. This ready-to-use paste of sky blue colour is suitable for application to the cut ends with a brush.

6. Bordeaux Paint

Bordeaux paint is applied for curing the deep and deformed wounds caused by cankers, gummosis, collar rot/root rot and to wounds resulting from surgical removal of crown gall or hairy roots from the stems of infected plants during dormant period. It is a suitable recommendation in areas or seasons which receive heavy downpour. It protects the treated part of the host against water for longer period, thus ensuring rapid healing of wounds.

Preparation of Bordeaux paint

To prepare the Bordeaux paint use monohydrated copper sulphate (1 kg), hydrated lime dust (2 kg) and boiled linseed oil (3 litres). Monohydrated copper sulphate can be easily made by heating the copper sulphate crystals on an iron sheet or in a frying pan till they crumble into a white amorphous form which is ground into a fine powder. Monohydrated copper sulphate powder can then be mixed thoroughly with lime dust and homogenised with 3 litres of boiled but cooled linseed oil to make a thick paste. This paste may be stored in a glass jar or other suitable non-metallic vessel for future use. It can be applied to the plant parts with the help of a brush. In Mango Bordeaux paste can be used after pruning if plant is infected with twig die-back disease. Bordeaux mixture can be sprayed after fruit set if area is affected with black tip disorder.

7. Important Mango Hybrids and their Parentage

S.No	Hybrid	Research Station	Parentage
1.	Mallika	IARI, New Delhi	Neelum x Dashehari.
2.	Amarapali	IARI, New Delhi	Dashehari x Neelum.
3.	Au-Rumani	F.R.S., Kodur	Rumani x Mulgoa.
4.	Manjeera	F.R.S., Sangareddy	Rumani x Neelum.
5.	PKM-1	H.R.S., Periyakulam	Chinnaswarnarekha x Neelum.
6.	PKM-2	H.R.S., Periyakulam	Neelum x Mulgoa.
7.	Ratna	F.R.S., Vengurla	Neelum x Alphonso.
8.	Sindhu	F.R.S., Vengurla	Ratna x Alphonso.
9.	Alfazli	F.R.S., Vengurla	Alphonso x Fazli.
10.	Sundar Langra	F.R.S., Vengurla	Sardar Pasand x Langra
11.	Arka Aruna	IIHR Bangalore.	Banganapalli x Alphonso
12.	Arka Puneet	IIHR Bangalore.	Alphonso x Banganapalli
13.	Arka Anmol	IIHR Bangalore.	Alphonso x Janardhan Pasand.
14.	Arka Neelkiran	IIHR Bangalore.	Alphonso x Neelum.
15	Pusa Arunima	IARI, New Delhi	Amrapali x Sensation
16	CISH M1	CISH, Lucknow	Amrapali x Janardan Pasand
17	CISH M2	CISH, Lucknow	Dasheri X Chousa
18	Nileshan	ARS, Paria , Gujarat	Neelum x Beneshan
19	Neelphonso	ARS, Paria, Gujarat	Neelum x Alphonso
20	Nileshwari	ARS, Paria, Gujarat	Neelum x Dasheri
21	Sonpari	ARS, Paria, Gujarat	Alphonso x Beneshan
22	Neelgoa	ARS, Ananat-rajupeth, A.P.	Neelum x Mulgoa

8. Technology Sources

The major sources for technology, as well as quality planting material are:

- Central Institute for Sub-tropical Horticulture, P.O. Kakori, Lucknow-226002, Uttar Pradesh, Tel (0522)-2841022/1023.
- Indian Institute of Horticultural Research, Hessarghatta, Bangalore-560089, Karnataka, Tel (080)-28466471/6353.
- Indian Agricultural Research Institute, Pusa, New Delhi-110012.
- Narendra Deva University of Agriculture & Technology, Kumarganj,Faizabad-224229, Uttar Pradesh, Tel (05270)- 2262097/ 2161.
- Acharya NG Ranga Agricultural University, Rajendra Nagar, Hyderabad-500030, Andhra Pradesh, Tel (040)-24015078.
- University of Agricultural Sciences, Dharwad-580005, Karnataka, Tel (0836)-2447783.
- Mahatma Phule Krishi Vidyapeeth, Rahuri-413722, Maharashtra, Tel (02426) 2243208.
- Dr. Balasaheb Sawant Konkan Krishi Vidyapeeth, Dapoli District, Ratnagiri-415712, Maharashtra, Tel (02358)-2282064.
- Directorate of Horticulture, of all the states.
- Central Horticultural Experiment Station, Aiginia, Bhubaneswar-Orissa.Tel (0674)-2471712.
- Horticulture and Agro-forestry Research Programme, Plandu, Ranchi, Jharkhand.Tel(0651)-2260207.
- Indira Gandhi Krishi Vishwa Vidyalaya, Krishaknagar, Raipur, Chattisgarh.
- Jawahallal Neharu Krishi Vishwa Vidyalaya,Adhartal, Jabalpur M.P.
- Govind Ballabh Pant University of Agriculture and Technology, Pantnagar, Udhamsinghnagar, Uttaranchal.
- Orissa University of Agriculture and Technology, Bhubaneswar, Orissa.

9. Cost of Mango Cultivation (Rs. per acre)

S. No.		Component	
Expenditure (Rs.)			
1.	Cultivation Expenses		
	(i)	Cost of planting material	2,000
	(ii)	Manures & fertilizers	5,000
	(iii)	Insecticides & pesticides	2,000
	(iv)	Cost of Labour	8,400
	(v)	Others, (Power)	3,600
		Subtotal	21,000
2.	Irrigation		
	(i)	Tube-well/submersible pump	45,000
	(ii)	Cost of Pipeline	-
	(iii)	Others,	-
		Subtotal	45,000
3.	Cost of Drip/Sprinkler		25,000
4.	Infrastructure		
	(i)	Store & pump house	15,000
	(ii)	Labour room	5,000
	(iii)	Agriculture Equipments	5,400
		Subtotal	25,400
5.	Land Development		
	(i)	Soil Leveling	4,000
	(ii)	Fencing	29,600
		Subtotal	33,600
	Grand Total		**1,50,000**

Source : Nation Horticulture Board

❑❑❑

23

Bibliography

Ahmad, S. and Ahmad, N. (1960). Mango receips. *Punjab Fruit Journal* 23: 169-172.

AMIA, (2006). Australian Mango Industry Association Ltd. Website accessed 5th June, 2007. *http://www. mangoes.net.au/general/ industry_profile/.*

Anand, J.C. and Johar, D.S. (1958). Improved method for the commercial preservation or raw mango slices. *Food Science (India)* 7: 150- 157.

Anonymous (1990). Improved cultivation practices for horticultural crops. Published by Directorate of Extension, UAS, Bangalore.

Anonymous (1986). Package of Practices for various fruit crops. Indian Agricultural Research Institute. New Delhi.

Anonymous (1990). Mango in India: Production, preservation and processing. CFTRI. Mysore, India, 1990: 99. 20.

Anonymous (1983). Mango cultivation , extension Bull. 9. IIHR, Bangalore.

Atwal, A.S. (1963) Insects pests of mango and their control. *Punjab Horticultural Journal.* 3 (2-4) : 325-358

Bajpai, P.N. and Shukla , H.S. (1978). Combating mango malformation through exogenous application of NAA. *Plant Science,* 10 : 135-137.

Baldry, J. Bregag, G.R. Cayagill, I.C. Cook, R.D., Ferber, C.E.M. and Lalitha, K. (1976). Alternative methods of processing mangoes. *Indian Food Packer.* 30 : 56-52

Bhaargava, B.S. Sharma and Chadha, K.L. (1988). Leaf nutrient guide for fruit and plantation crops. *Fertilizer News*. 33, 21-29

Bhatnagar, H.C. and Subramanyam, H. (1973). Some aspects of presentation, processing and export of mango and its products. *Indian Food packer*. 27. 33-52

Bhullar, J.S.(1982). Ripening of Langra mangoes with ethrel and calcium carbide. *Progressive Horticulture*. 14(1) : 71-72.

Butani, D.K. (1979). Insects and Fruits. Delhi, India: Periodical Expert Book Agency. New Delhi

CAB International (2004). Crop Protection Compendium, (2004 Edition). Wallingford, UK,

C.B.S. Rajpoot and Singh, J.N. (1983). Effect of urea and GA3 sprays on the growth flowering and fruiting characters of mango. *Progressive Horticulture*. 15. 174- 177.

Chauhan, B.K. and Dabas, H.K. (1997). NHB production year book. National Horticulture Board. New Delhi: 129.

Chandra, A (1988). Note on control of pre harvest fruit drop in mango. *Current Agriculture*. 12. 83-84

Chadha, K.L. Pal R.N. Prakash, O. , Tandon, P.L. and Singh, H. (1979). Studies on mango malformation. Its causes and control. *Indian Journal of Horticulture*. 36: 359-368.

Chattopadhyay, N.C. and Nandi, B. (1977b). Chemical control of malformation of mango. *Indian Agriculturist*. 21: 217-221.

Chaudhary,S.K. and Nautiyal, J.P. (1985). Effect of deficiency of calcium. Magnesium and sulphur on the uptake of other nutrients in mango. *Acta Horticulture*. 231, 296-300.

Dahshan, D.I. (1987) Physiological studies on malformation of mango panicles. *Annals of Agricultural Science*. 32. 565 – 575

Dasgupta, S.N. and Sen, C. (1958). On the prevention of mango necrosis, *Current Science*. 27 : 446 – 447.

Dathathry Pillay, T.N. (1954). Mango in the orchard. The Mango. A Souvenir. Hyderabad. Department of Agriculture.

Fasih, M. and Srivastava, R.P. (1990). Parasites and predators of insect pests of mango. *Int. Pest Control*. 32:39-41.

Flint, M.L. and Van Den Bosch, R. (1981). Introduction to Pest Management. Plenum Press, New York.

Gupta, B.P. and Singh, Y.P. (1988). Mango scale insects occurrence in western Uttar Pradesh and their control. *Progressive Horticulture*. 20 (3-4) : 357-361).

Jhala, R.C., Patel, Z. and Shah, A. (1987). Studies on the relative occurrence of the leaf-gall midge (*Procontarinia matteiana* Keiffer and Cecconi) on different varieties of mango in South Gujarat, India. *Trop. Pest Manage.* 33:277-279.

Jothi D.B., P.L. Tandon and A. Verghese (1994). Hot water immersion as a quarantine treatment for Indian mangoes infested with the oriental fruit fly, *Bactrocera dorsalis* (Hendel) (Diptera:Tephritidae) *FAO Plant Protection Bulletin* 42(3):158- 159

Karra S.K.and Tandon, D.K. (1982). Regulation of ripening of mango cv. Mallika. *Inidan Journal of Horticulture* 39(2) : 155-159.

Kausar, A.G. (1959).Malformation in inflorescence of mango, *Punjab Fruit journal*, 22 : 19-21.

Khurdiya, D.S. and Roy, Susanta K. (1986). Studies on repining and canning of mangoes, *Indian Food packers* 40 (1) : 45 -48.

Kumar, S and Nautiyal. J.P. (1979). Foliar sampling technique in mango. *Punjab Horticultural Journal* 19. 10-15.

Ledger. S.N. (1991). Post harvest handling of mangoes. *Acta Horticulture* 291 : 479-488.

Majumdar, P.K. and Sharma, D.K. (1990). In T.K. Bose (ed) Fruits : Tropical and Sub-tropical , Naya Prakash, Calcutta, pp 1 – 62.

Majumdar, P.K. Shrarma, D.K. Singh. M.P. and Singh , R.N. (1976). Improve productivity of malformed mango trees. *Indian Horticulture* 20 : 7-8

Mallik, P.C. and De, B.N., (1953). Manures and manuring of mango and the economics of mango culture. *Indian Journal of Agricultural Sciences*, 22 : 251.

Mallik , P.C. and Singh , D.L. (1959). Deficiency symptoms in mangoes due to the absence of trace elements. *Indian Journal of Horticulture*, 19 : 228-231.

Misra A.K. (1992). Integrated management of mango diseases. (Abs.) Indian physiopathology 43 & 44 (Suppl.) LXIV.

Mohyuddin, A.I. and Mahmood, R. (1993). Integrated control of mango pests in Pakistan. *Acta Hortic.* 341:467-483.

Mukunda, G.K.and Narayana,Reddy , M.A. (1998).Soil and climate In: Mango cultivation. R. P. Srivastava (editor) International Book Distribution Co. Lucknow.

Pal, R.N. Chadha, K.L. and Rao, M.R.K. (1984). Effect of different plant growth regulators and other chemicals on flowering behaviour of mango. *Indian journal of Horticulture*. 41, 8-15.

Pal, R.N. and Chadha, K.L. (1980). Black tip disorder of mango – A review, *Punjab Horticulture Journal* 3: 221-25.

Pandey, R.M., Rathore, D.S. and Singh, R.N. (1974). Hormonal regulation of mango malformation. *Current Science* 43 : 694 – 695.

Pandey, R.M. , Singh R.N. and Sharma, V.K. (1974). Leaf scorch in mango, a new problem. *Indian Horticulture* (April – June) 19: 7-8

Pandey, R.M and Om Prakash (1998). Mango disorders, causes and management. In: Mango cultivation. R. P. Srivastava (editor) International Book Distribution Co. Lucknow.

Pandey, S.N.(1984). International Check list of Mango Cultivars. pp. 284. Indian Agricultural Research Institute. New Delhi

Pandey, S.N. (1998). Mango cultivars. In: Mango cultivation. R. P. Srivastava (editor) International Book Distribution Co. Lucknow.

Patel, R.K., Patel, S.R. and Shah, S. (1977). Biology of mango hopper, *Amritodus atkinsoni*(Leth.) (Jassidae: Hemiptera) in South Gujarat. *Indian J. Entomol.* 37:150-153.

Peña, J.E. and Mohyuddin, A.I. (1997). Insect pests. In: Litz, R.E. (ed.). The Mango:Botany, Production and Uses. Wallingford, UK: CAB International, pp. 327-362.

Prakash, Om, Misra, A.K. and Ram Kishun (1997). Some threatening diseases of mango and their management, Management of Threatening Plant Diseases of National Importance (Ed. V.P. Agnihotri et al.). Malhotra publishing House, New Delhi : 179-205.

Prakash , Om and Misra. A.K. (1993). Fungal Diseases of Sub-tropical horticultural fruit crops. *Advance in Horticulture*, (Ed. K.L. Chadha). Malhotra Publishing House, New Delhi (India)

Rahman, K.A. (1940).Important insect pests of the mango and how to control them. *Punjab Fruit J.* 3.

Ranga Reddy, K., (1954). Commercial mango orchards. The Mango. A Souvenir. Department of Agriculture, Hyderabad (India)

Rao, M.M. (1998).The problem of alternate bearing in mango: causes and management In: Mango cultivation. R. P. Srivastava (editor) International Book Distribution Co. Lucknow

Roy, S.K and C.K.Narayana (1998). Post harvest technology in mango In: Mango cultivation. R. P. Srivastava (editor). International Book Distribution Co. Lucknow

Samra, J.S. and Arora Y.K (1998). Nutrition in mango. In: Mango cultivation. R. P. Srivastava (editor) International Book Distribution Co. Lucknow

Sen, P.K., Roy, P.K. and De, B.N. (1947). Hunger signs on mango, *Indian Journal of Horticulture*. 5 : 35.

Sen, P.K. (1943). The bearing problem of mango and how to control it. *Indian journal of Horticulture* 1, 48-71

Shukla, R.P. and Prasad, V.G. (1984). Evaluation of some insecticides against mango hopper, Idioscopus clypealis (Lethierry) (Homoptera: Cicadellidae). *Indian J. Agric. Sci.* 54:677-681.

Shukla, R.P. and Tandon, P.L. (1985). Bio-ecology and management of the mango weevil, Sternochetus mangiferae (Fabricius) (Coleoptera: Curculionidae). *Int. J. Trop. Agric.* 3:293-303.

Srivastava, H.C. (1967). Grading, Storage and Marketing the Mango: A Hand Book. ICAR, New Delhi.

Srivastava, R.P. Pests of mango (1998). In: Mango cultivation. R. P. Srivastava (editor) International Book Distribution Co. Lucknow

Srivastava, R.P. (1987). Recent advances on key pests of mango and their management. In : Recent Advance in Entomology. (Ed. Mathur et al.) Gopal Prakashan, Kanpur, pp. 528.

Srivastava, R.P. and Tandon, P.L. (1983). Field evaluation of some insecticides for the control of mango blossom midge. *Erosomyia indica. Pesticides* 17 (3) :21

Srivastava, R.P. and Tandon, P.L. (1986). Natural occurrence of two entomogenous fungi pathogenic to mango hopper, *Idioscopus clypealis* Leth. *Indian J. Plant Pathol.* 4:121-123

Singh, D. (1976). The Mango: A Handbook, ICAR, New Delhi, pp. 17-18.

Singh, G. (1975). Important insect pests of fruits and their control. *Indian farmers digest* 8:46.

Singh, G. (1993). Insect pests of mango. In advances in Horticulture (Fruit Crops) Vol – III Eds. Chadha, K.L. and Pareek, O.P. p. 1488.

Singh, L.B., (1960). The Mango: Botany, Cultivation and Utilization. Leonard Hill. London.

Singh, Ranvir and Singh, C. P. (1998). Planting and care, propagation, root stock and irrigation in mango. In: Mango cultivation. R. P. Srivastava (editor) International Book Distribution Co. Lucknow.

Singh, R.N. (1978). Mango. Indian Council of Agricultural Research, New Delhi. pp.39-55

Singh, R.S. (1968). Plant diseases. Oxford and IBH Publishing Co., New Delhi.

Singh, S.M. (1956). Control of mango malformation disease. *Science and Culture* 21 (9) : 532 -533

Sujatha, A., M.S.V. Chalam, K. Bheem Rao and B.Gautam (2007). Management of mango fruit borer *Deanoles albizonalis* (Hanpson). *Technical Bulletin*. Agricultural Research Station, A.N.G.R. Agricultural University Ambajipeta and Agricultural and Processed Food Products Export Development Authority (APEDA).pp-4

Tandon, P.L. (1991). Problems and prospects of insect pest management in fruit trees. In *Trends in Agricultural Pest management* (G.S. Dhaliwal and R. Arora, eds.). Commonwealth publishers, New Delhi.

Tandon, P.L. and A. Verghese (1985). World list of insect, mite and other pests of mango. *Technical Document No.5*, IIHR, Bangalore, 22p.

Teoatia, S.S. Singh, R.D. and Maurya, V.N. (1968). Studies on maturity standards for *Mangifera Indica* L. Var. Langra, *Indian Journal of Horticulture* 25 : 24-30.

Thimmappaiah and Suman, C.L. (1987). Sex ratio in relation to fruit set and fruit yield in mango. *Punjab hort.* J.27: 8-11

Thompson, A.K. (1996). Post Harvest technology of fruits and vegetables. Blackwell Science Inc., Cambridge, pp. 26.

Uthaiah, B.C., Indiresh, K.M., Hussain, I.S.A., Rao, K.B. and Hanummaiah, H. (1988). Flower and sex variation in mango varieties under coastal Karnataka. *Prog. Hort.* 20: 120-123

Veeresh, G.K. (1989). Pest problems in mango - world situation. *Acta Hortic.* 231:551-565.

Verghese, A, and Rao, G.S.P. (1983). Determination of relevant critical stages for the management of the mango hopper. *Idioscopus `clypealis* Leth.*Indian journal of Horticulture* 44 (3-4) : 280-283.

❏❏❏

24

Colour Plates

Colour Plates

(A) Glimpses of Insects and their Damage

Ceroplastes sp.

Rastrococcus sp.

Chloropulvanaria sp.

Aspidiotus sp.

Scale Insects / Mealy Bugs

Shoot borer

Slug caterpillers

Mango mealy bug

Mango leaf cutting weevils & leaf damage

Egg laying female & fruit damage by fruit fly

Leaf miner and its larva

Termite infestation in mango

Stone weevil adult, larva & damaged stone

Fruit borer larva & its damage

Leaf webber damage in mango

Shoot gall made by psylla

Larva & adults of fruit sucking moths

Gallery made by bark eating cater piller

Black fly

Leaf galls

Hoppers & their damage

Fruit damaged by thrips

Stem borer larva & its damage

(B) Disease Symptoms in Mango

Anthracnose of mango fruits

Anthracnose of mango leaves

Die back disease of mango

Powdery mildew on mango inflorescence

Root rot and damping off of mango seedlings

Stem end rot of mango fruit

Black rot of mango fruit

Red rust disease on mango leaves

Black banded disease of mango

Scab disease of mango seedlings

Bacterial black spot
(canker) on fruits

Bacterial black spot
(canker) on leaves

Black tip disorder of mango fruit

Vegetative malformation of mango

Gummosis of mango

Sooty mold on mango leaves

(C) Rejuvenation Technique

Old orchard with inaccessible canopy growth

Cutting of inaccessible branches

Pasting of cut portion

Vigoursly pruned tree

Sprouting in pruned tree

Shoots ready for grafting

Growth of grafted shoots

Removal of ungrafted shoots

Canopy growth of rejuvenated plant

Flowering in rejuvenated plant

Fruiting in rejuvenated plant

Intercropping in rejuvenated orchard